Modeling, Optimization, and Control of Zinc Hydrometallurgical Purification Process

Modeling, Optimization, and Control of Zinc Hydrometallurgical Purification Process

Chunhua Yang
Central South University
School of Automation
Changsha, Hunan, China

Bei Sun
Central South University
School of Automation
Changsha, Hunan, China

Series editor
Quan Min Zhu

ELSEVIER

ACADEMIC PRESS
An imprint of Elsevier

Academic Press is an imprint of Elsevier
125 London Wall, London EC2Y 5AS, United Kingdom
525 B Street, Suite 1650, San Diego, CA 92101, United States
50 Hampshire Street, 5th Floor, Cambridge, MA 02139, United States
The Boulevard, Langford Lane, Kidlington, Oxford OX5 1GB, United Kingdom

Library of Congress Cataloging-in-Publication Data
A catalog record for this book is available from the Library of Congress

British Library Cataloguing-in-Publication Data
A catalogue record for this book is available from the British Library

ISBN: 978-0-12-819592-5

For information on all Academic Press publications
visit our website at https://www.elsevier.com/books-and-journals

Publisher: Mara Conner
Acquisitions Editor: Sonnini R. Yura
Editorial Project Manager: Andrea Gallego Ortiz
Production Project Manager: Prasanna Kalyanaraman
Designer: Miles Hitchen

Typeset by VTeX

Contents

Part II
Modeling and optimal control of the copper removal process

Part III
Modeling and optimal control of the cobalt removal process

About the authors

Chunhua Yang is currently a full professor, Dean of the School of Automation, Central South University. She is a winner of the National Science Fund for Distinguished Young Scholars of China. Prof. Chunhua Yang received her B.Eng. Degree and her M.Eng. Degree in Automatic Control Engineering and her Ph.D. in Control Science and Engineering from Central South University, China in 1985, 1988, and 2002 respectively. She worked as a visiting professor with the Department of Electrical Engineering, Catholic University of Leuven, Belgium from 1999 to 2001. From 2009 to 2010, she was a Senior Visiting Scholar at the University of Western Ontario, Canada. Her research interests mainly include control and optimization of complex industrial processes, and smart manufacturing of process industries. She has received funding in more than 20 scientific research projects, including the Major Program, the Key Program, and Distinguished Young Scholars Fund of the National Nature Science Fund of China, the National High Technology Research and Development Program of China, the National Key Technology Research and Development Program of the Ministry of Science and Technology of China, etc. She has won the Second Prize of National Science and Technology Progress for 4 times. Her publications include 3 monographs, more than 300 papers with 160 articles in international peer-reviewed journals, such as IEEE Trans. on Industrial Electronics, IEEE Trans. on Fuzzy Systems, Journal of Process Control, Control Engineering Practice and so on. She is a vice-chair of the IFAC Technical Committee on Mining, Mineral and Metal Processing, and a Fellow of the China Association of Automation.

Bei Sun is currently an associate professor at the School of Automation, Central South University. He obtained his PhD degree in 2015 from Central South University. He was a jointly supervised PhD student of New York University, USA from 2012 to 2014. From December 2015 to September 2018, he was a lecturer at the School of Information Science and Engineering, Central South University. His research interests include process modeling, identification and control, reinforcement learning, and smart and optimal manufacturing of process industries. He is a recipient of the First Prize of Science and Technology of the Nonferrous Metals Society of China (2016), the IFAC Young Author Award (finalist, 2018), and the First Prize of Technical Invention of the Ministry of Education of China (2019). He is a member of the IFAC Technical Committee on Mining, Mineral and Metal Processing, and the secretary of the Academic Committee on Automation of the Nonferrous Metals Society of China.

Preface

Zinc hydrometallurgy process is a typical metallurgical process, which involves the transformation from low-grade raw minerals to very high-grade metal products. For the zinc hydrometallurgy process, the transformation from zinc concentrate ore to pure zinc metal is gradually realized through interconnected unit processes, mainly including roasting, leaching, purification, and electrowinning, where various physicochemical reactions take place. Understanding the complex process dynamics and making appropriate optimization or control decisions according to the feeding and reaction conditions are essential in guaranteeing the product quality and reducing the production cost. In contrast, manual operation is subjective and relies on individual experience, which can lead to large fluctuations and low utilization rates of valuable mineral sources. With the shrinking of high-grade mineral resources, more fierce global competition, and more stringent environmental regulations, the necessity of applying process automation and control technologies to achieve green and efficient production has been fully realized by the metallurgical industry. In the era of the Fourth Industrial Revolution, the demand of smart manufacturing will promote the development of process automation and control technologies to a higher level.

In the past 30 years, the authors' research group is dedicated to research on the modeling, control, and optimization for the nonferrous metallurgical processes. By insisting the tradition of conducting substantive work from the perspective of both theory and practice, the research group extracts scientific and engineering problems from practical production process and explore effective solutions. These solutions have played a positive role in promoting the automation level and production efficiency in nonferrous metallurgical plants. This book summarizes the authors' previous work on the modeling and operational optimization of the solution purification process of a real zinc hydrometallurgical plant and also presents some recent theoretical studies on the controller design of metallurgical reactors. By taking the purification process as the research object, this book provides an overview of how to combine the knowledge of related disciplines to realize the description of process dynamics and the optimization of operational variables for practitioners as well as graduate or senior undergraduate students. In particular, this book emphasizes the following topics: (1) how to apply metallurgical knowledge to study the reaction mechanism,

(2) how to describe and model the process dynamics, (3) how to design an optimization framework based on an analysis of the characteristics of the process flow, (4) how to estimate key process indicators by reconstructing online measurements, and (5) how to adjust manipulated variables using machine learning or control approaches.

The book is divided into four parts. Part I (Chapters 1 and 2) presents the research background. In Chapter 1, typical zinc hydrometallurgy technologies, including the roasting-leaching-electrowinning technology and the atmospheric direct leaching zinc hydrometallurgy technology, are introduced. The functions of each unit process are introduced, in which the solution purification process is introduced in more detail. In Chapter 2, the challenges in the modeling and optimal control of the solution purification process are first analyzed. A comprehensive and in-depth discussion of process modeling is presented. Then, an integrated modeling framework and a cooperative optimization and control framework are proposed. Part II (Chapters 3 through 5) presents the modeling, optimization, and control of the copper removal process, which is a unit process of the solution purification process. First, in Chapter 3, the copper removal process is abstracted and modeled as a competitive-consecutive reaction system. Then, Chapters 4 and 5 provide approaches to estimate and adjust the zinc powder dosage of each reactor. Part III (Chapters 6 through 8) presents the modeling, optimization, and control of the cobalt removal process, which is another unit process of the solution purification process. The relationship between the oxidation–reduction potential and the reaction rate is studied in Chapter 6, based on which the kinetic model of the cobalt removal process is built. Then, according to the characteristics of the cobalt removal process flow, i.e., the process is composed of multiple cascaded reactors, a "gradient optimization" strategy is proposed in Chapter 7. Chapter 8 presents some theoretical studies on the control of the cobalt removal reactor under different working conditions. Part IV includes Chapters 9 and 10. Chapter 9 briefly introduces the development of control systems of the solution purification process. Chapter 10 summarizes the book and presents a discussion on some potential research directions.

Acknowledgments

The publication of this book would have been impossible without the support of many people. First, the authors wish to thank Prof. Weihua Gui, whose guidance and constructive suggestions inspired many ideas of this book. Second, the authors gratefully acknowledge the many current and previous project team members, especially Profs. Yonggang Li and Hongqiu Zhu, Drs. Bin Zhang, Tiebin Wu, and Mr. Tianshui Lin, for their effort on site and in the laboratory, selfless collaboration, and useful comments during the long-term research process. Third, many thanks go to the Elsevier team, particularly Acquisitions Editor Sonnini R. Yura, Project Manager Prasanna Kalyanaraman, Content Delivery Manager Andrea Gallego Ortiz, and Copyrights Coordinator Indhumathi Mani, for their professional assistance and patience during the publishing process. We are grateful to Prof. Fanbiao Li, Dr. Jie Han and all the anonymous reviewers who carefully reviewed the table of contents and the final version of this book. The authors are grateful for the support from the National Key R&D Program of China (Grant No. 2018YFB1701100), the Key Project of International Cooperation and Exchanges NSFC (Grant No. 6180206014), the Science Fund for Creative Research Groups of the National Natural Science Foundation of China (Grant No. 61621062), the National Natural Science Foundation of China (Grant Nos. 61988101, 61973321), and the 111 Project (B17048).

Chunhua Yang and Bei Sun
Central South University, Changsha
August, 2020

Part I

Background

Chapter 1

Introduction

Contents

1.1 Overview

Zinc is an important and commonly used nonferrous metal [1]. The annual consumption of zinc is ranked third among all nonferrous metals, after only copper and aluminum. The chemical symbol of zinc is Zn, which has its roots in Latin, "Zincum," meaning white thin layer or white deposit. The color of zinc metal is silver-white mixed with light blue-gray. The atomic weight and atomicity of zinc are 65.409 and 2, respectively [2]. The densities of solid zinc and liquid zinc are 7.14 g/cm^3 and 6.62 g/cm^3, respectively. In the periodic table of elements, zinc is the 30th element, which is one of the forth period elements, or group 12 elements. In a zinc atom, the first, second, and third shells are filled with electrons. The forth shell has two electrons located on the 4s orbital. Therefore, all the electron orbitals of zinc are fully filled. Due to its special atomic structure, zinc is chemically active, and has excellent mechanical and physicochemical properties, e.g., malleability, abrasive resistance, low melting point and good melt liquidity, anticorrosion property, etc. Owing to the above properties, zinc has been widely used in various fields contributing to the national economy, especially in the auto industry, construction industry, shipping industry, and light industry [3].

Zinc seldom exists in the elemental form in nature. Instead, it normally exists in combination with other base metals such as copper and lead in various kinds of zinc ores, e.g., sphalerite, smithsonite, zincite, willemite, and hemimorphite, most of which are zinc sulfide ores. Zinc has been used with other metals in the form of alloy for several thousand years, e.g., brass, which is an alloy of zinc and copper, and bronze, which is an alloy of zinc and tin [4]. Before the invention of hydrometallurgy technologies, metals were produced using high-temperature processes. As the boiling point of zinc (907°C) is lower than that of most common metals (copper, aluminum, iron, lead), it is converted into zinc

https://doi.org/10.1016/B978-0-12-819592-5.00010-7

vapor and escapes before these associated metals are obtained. Therefore, zinc can hardly be observed. The production of elemental zinc became possible after the invention of gas condensation technology. Since then, the production of zinc has started in many countries (see [3] for more details), and was included in the list of elements as a metal element by Antoine-Laurent de Lavoisier in 1789.

Essentially, the production of pure zinc metal involves the extraction of special high-grade (SHG, 99.995% purity) metallic zinc from zinc ores. Zinc production technologies are designed according to the physicochemical properties of zinc, which usually contains multiple stages. Each stage includes different tasks and produces a certain intermediate product. Zinc ore is gradually transferred into zinc metal via these stages. Pyrometallurgy and hydrometallurgy are the two main approaches to produce zinc. Zinc pyrometallurgy technology can be traced back to the 1020s. It had long been the main zinc production technology before the invention of zinc hydrometallurgy technology, which was commercialized in the 1910s. Compared with zinc hydrometallurgy, zinc pyrometallurgy has disadvantages like more serious environmental pollution and higher energy consumption. Therefore, most zinc pyrometallurgy technologies are retired, except imperial smelting process (ISP) technology, electric furnace zinc smelting, and some other improved zinc pyrometallurgy technologies. Nowadays, zinc hydrometallurgy is the main approach in zinc production [5][6]; more than 85% of the world's zinc is produced by zinc hydrometallurgy.

In zinc hydrometallurgy plants, the objective of routine operation is to keep each stage running under appropriate conditions, thus to:

- Guarantee the quality of both intermediate and final products.
- Sustain the stability of the process.
- Increase the recovery rate of zinc and all the associated valuable metal/non-metal elements.
- Minimize the overall production cost and environment effects.

To achieve this aim, numerous researchers from the metallurgy community have conducted studies on the reaction mechanism of the zinc hydrometallurgy process via laboratory/pilot plant-scale experiments or analyzed the resultant samples from real plants. These studies mainly include:

- Deduction of the reaction type [8].
- Construction of process models [10].
- Investigation of the influence of process variables on the technical indices [9].
- Determination of suitable reaction conditions [7].

These results can improve our understanding of process dynamics, promoting the improvement and innovation of technology and providing guidance for the daily operation and optimal control.

With the shrinking of high-grade zinc ore resources and fierce global competition among zinc smelting enterprises, zinc concentrates of a zinc smelting plant are purchased from different mines to optimize the supply of raw materials.

(a) Zinc concentrate (b) Different types of zinc concentrates

(c) Zinc ingots

FIGURE 1.1 Zinc ore and zinc ingots.

These different concentrates have different grades and physicochemical properties, which can be recognized from the color of the concentrate in different proportioning bins, as shown in Fig. 1.1. Therefore, the composition of feeding material is intricate and tends to suffer from fluctuations. This gives rise to more complex reactions and variation of process dynamics. However, the quality of the final zinc product must meet the standards. More specifically, metal impurities directly affect the mechanical properties of the zinc product, which should be removed to safe levels in the solution purification stage. Therefore, the understanding of process dynamics and reasonable determination of manipulated variables of the solution purification process are of vital importance in zinc hydrometallurgy, which is the main concern of this book.

1.2 Zinc hydrometallurgy technologies

Zinc hydrometallurgy technology involves the use of aqueous solutions to recover zinc from zinc ores [11]. According to the literature, the first published research on zinc hydrometallurgy dates back to 1871. Since then, zinc hydrometallurgy technology has undergone three stages of development.

(i) The first stage was from the late 19th century to the middle of the 1910s. During this period, the zinc hydrometallurgy technology was in the laboratory stage. The main activities in this stage were the experimental studies of subprocesses of zinc hydrometallurgy, e.g., in 1871, de Boisbaudran observed that cobalt cementation occurred if metal ions easily reducible by zinc (e.g., Cu^{2+}) were presented in the solution [8].

(ii) The second stage was from the middle of the 1910s to the 1960s. During this stage, the zinc hydrometallurgy technology became more mature and was commercialized. In the First World War (1914–1918), the demand of cartridge brass accelerated the development of large-scale zinc hydrometallurgy production technology. The hydrometallurgical zinc electrowinning process, or roast-leach-electrowinning (RLE) zinc production technology, was put into practice in North America in the middle of the 1910s, including Cominco in Trail, British Columbia and Anaconda in Montana. Now, the RLE technology and its varieties account for more than 85% of the total zinc production [12].

(iii) The third stage is from 1960s to now. In this stage, the zinc hydrometallurgy technology keeps improving, and the yield of zinc hydrometallurgy gradually surpasses that of zinc pyrometallurgy. The recovery of associated valuable metal/nonmetal elements is also considered in the technological improvements. A milestone was the invention of a series of iron removal technologies in the 1960s, e.g., the goethite process, hemmatite process, and jarosite process [13]. The separation of zinc and its associated elements increased the quality of the final zinc product and decreased the content of zinc in residues. Therefore, the overall zinc hydrometallurgy process contains not only the main steps to produce zinc metal, but also other secondary steps to produce by-products, e.g., residues of iron, copper, cadmium, cobalt, and sulfur, which are then further refined. In addition, the pressure leaching of zinc sulfide concentrates was developed in this stage. In the conventional RLE process, the zinc sulfide concentrate has to be roasted to zinc oxide before leaching, so technically, the RLE process is a combination of pyrometallurgy and hydrometallurgy. In the 1970s, research on the "direct leaching" technology started [14]. In direct leaching, zinc sulfide concentrate is directly processed to liberate zinc ions without roasting [15]. The overall process then becomes a fully hydrometallurgical one [16][17]. The atmospheric direct leaching zinc hydrometallurgy technology is the most advanced commercialized zinc hydrometallurgy technology at present.

Besides the evolution of zinc hydrometallurgy technology, driven by the development of information and communication technology (ICT), distributed control systems have been installed to monitor and control zinc hydrometallurgy plants. In recent years, with the global trend of smart and intelligent manufacturing as well as more strict requirements on the recovery rate, production efficiency, and product quality, more advanced process control approaches are being applied to zinc hydrometallurgy processes for stable and economical production.

1.2.1 Roasting-leaching-electrowinning zinc hydrometallurgy technology

The RLE process is the conventional zinc hydrometallurgy technology. It is generally composed of five major steps, i.e., roasting, leaching, solution purification, electrowinning, and casting. As shown in Fig. 1.2, the existing form of zinc experiences complex transitions from complex zinc ore to pure zinc metal. First, zinc concentrate is transformed to zinc oxide in the roasting process. Then, in the leaching process, the solid oxidized concentrate ores are treated in sulfuric acid solution in order to liberate the valuable metal ions from concentrate ores. Due to the impurity and heterogeneity of the concentrate ores, the other associated metallic ions in the concentrate ores are simultaneously dissolved into the acid solution. Therefore, the resulting leaching solution contains ions of impurity metals harmful to the electrowinning process in which the pure valuable metal is recovered by electrodeposition. The recovered zinc metal is then casted into final zinc products with different specifications, e.g., zinc ingot.

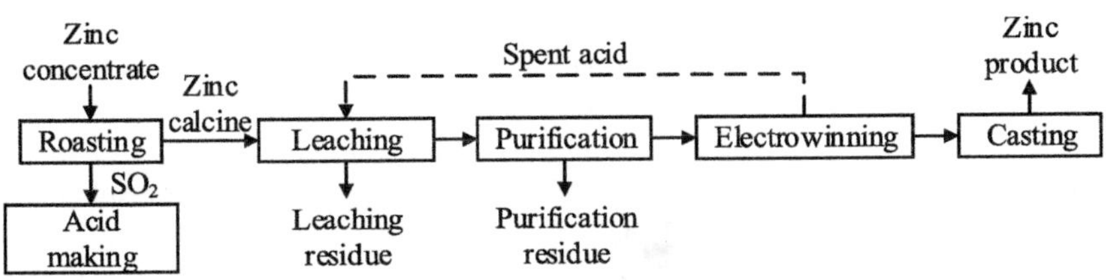

FIGURE 1.2 RLE process.

1.2.1.1 Roasting process

Zinc sulfide is the main content of zinc concentrate. It is insoluble in sulfide solution, and cannot be directly used as the feed of the leaching process. Therefore, as the first step of the RLE process, the task of roasting is to convert zinc concentrate to calcine. The calcine is mainly composed of zinc oxide, which is solvable in leaching solution. As shown in Fig. 1.3, the roasting is usually carried out within a roaster. The zinc concentrate is fed to the roaster via a thrower. Air is blown from the bottom of the roaster to conduct the following reactions with zinc sulfide:

$$ZnS + 1.5O_2 \rightarrow ZnO + SO_2,$$
$$ZnS + 2O_2 \rightarrow ZnSO_4,$$
$$2FeS_2 + 5.5O_2 \rightarrow Fe_2O_3 + 4SO_2,$$
$$ZnO + Fe_2O_3 \rightarrow ZnFe_2O_4.$$

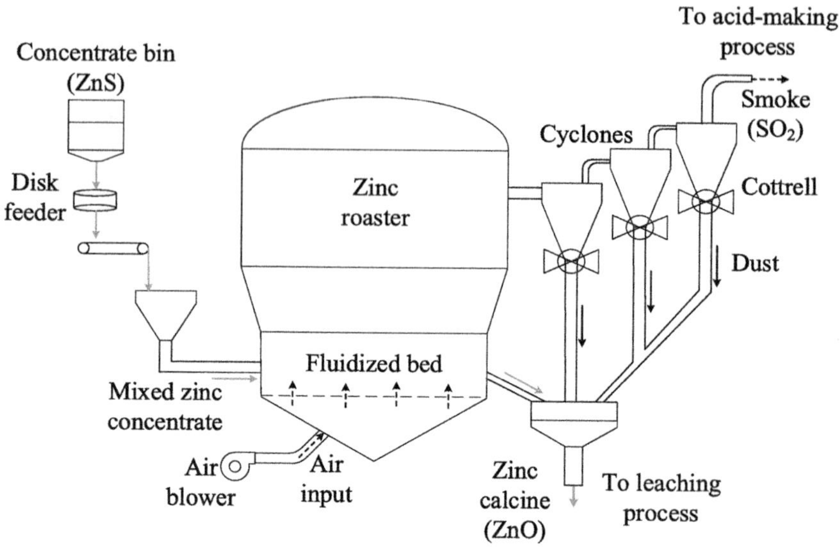

FIGURE 1.3 Roasting process.

The resultant SO_2 is delivered to the acid-making process. Calcine and roasting dust are collected and fed to the subsequent leaching process.

The technical indices of the roasting process include the content of soluble zinc and insoluble sulfide, which largely rely on the oxidation atmosphere or the temperature inside the roaster. The roasting temperature should be within reasonable range. As the above reactions are highly exothermic, the feeding rate and composition (i.e., content of zinc, lead, and silicon) of zinc concentrate directly affect the roasting temperature. Therefore, in practice, the roasting temperature is mainly controlled by adjusting the feeding rate of zinc concentrate. The blast volume, which also affects the roasting temperature, is usually fixed to maximum to guarantee high yield.

1.2.1.2 Leaching and purification processes

In the leaching process, dilute sulfide solution or spent acid from the electrowinning process is used as solvent to dissolve the zinc in calcine, such that the zinc ions are liberated and enter the solution. There are various types of leaching technologies, depending on the reaction conditions and number of stages. Each stage is composed of several leaching reactors. The leaching solution of the last stage is recycled to the first stage to form a leaching circuit (Fig. 1.4). For all the leaching technologies, the main chemical reaction involved is

$$ZnO + H_2SO_4 \rightarrow ZnSO_4 + H_2O.$$

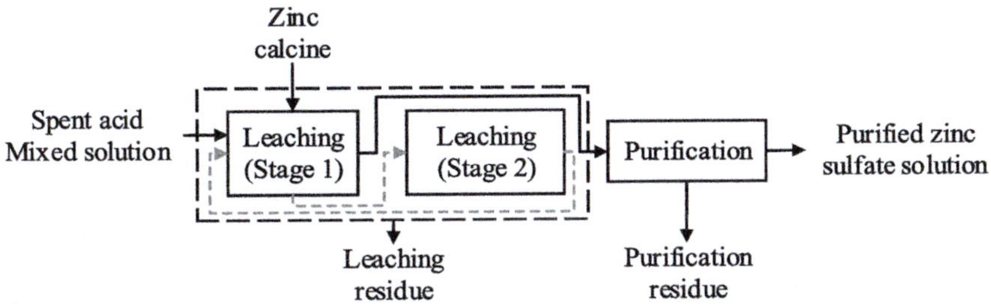

FIGURE 1.4 Leaching and purification process.

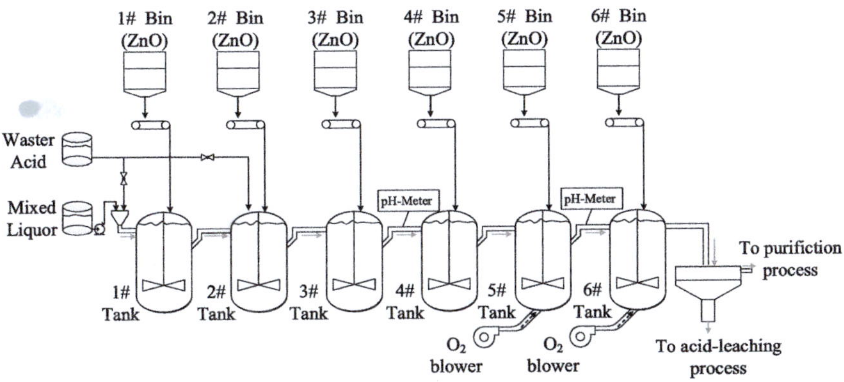

FIGURE 1.5 Leaching process.

The technical index of the leaching process is the zinc leaching rate. According to the thermodynamics and kinetics of the leaching process, the leaching rate is mainly determined by the concentration of hydro ions or the pH of each reactor. As shown in Figs. 1.4 and 1.5, leaching is the intersection of the solid-phase process (roasting) and the liquid-phase process (leaching, purification, electrowinning). In addition, spent acid from the electrowinning process and the mixed solution from various processes are recycled to the leaching process. The feeding conditions of the leaching process are uncertain and fluctuate frequently. Therefore, the pH value of each reactor must be controlled within appropriate ranges.

As zinc concentrate is not pure, zinc calcine obtained by roasting contains not only ZnO but also compounds of other elements, e.g., Fe, Si, and As. In the leaching process, besides the extraction of zinc ions, these impurity ions are also liberated and enter the leaching solution. Part of these impurities are deposited in the leaching process, e.g., Fe and Si. The precious metal residue is collected for further refining. The remaining impurities in the leaching solution, which have to be processed using specific technology, are removed in the subsequent purification process, which is introduced in more detail in Section 1.3. After leaching and purification, the zinc sulfate solution is of high purity.

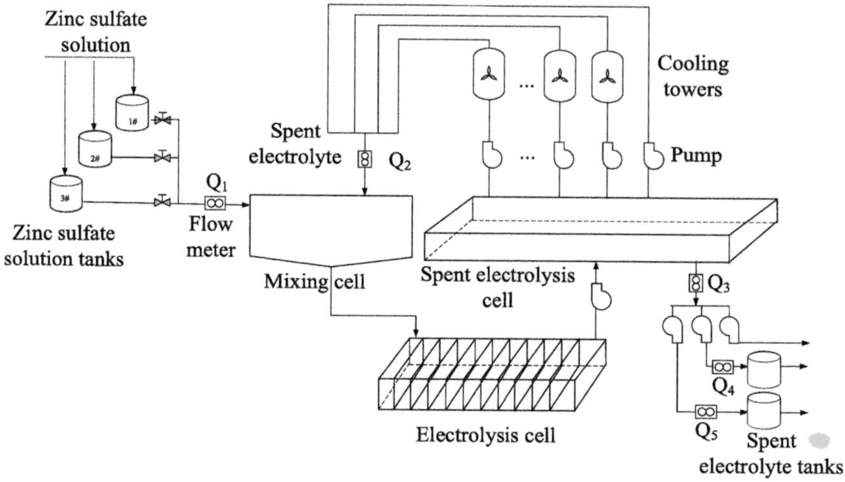

FIGURE 1.6 Electrowinning process.

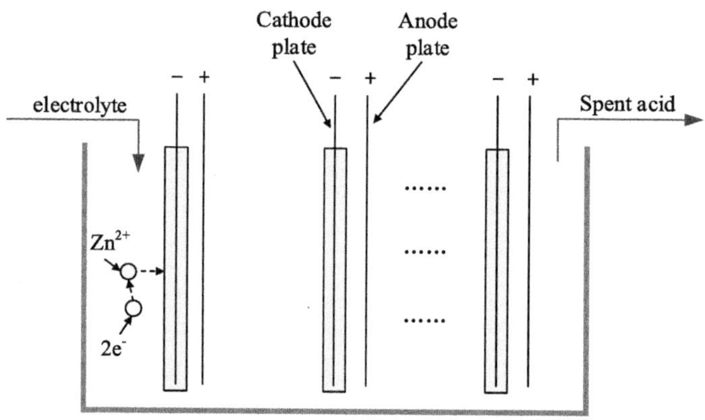

FIGURE 1.7 An electrolytic cell.

1.2.1.3 Electrowinning process

The zinc electrowinning process consists of several series of electrolytic cells which extract pure metal zinc from electrolyte (zinc sulfate solution) (Fig. 1.6). An electrolytic cell is composed of a cathode (aluminum plate) and an anode (plate of lead-silver alloy) (Fig. 1.7). By running a direct current through the cell, electrons are transferred between the anode and the cathode. A hydrolysis reaction takes place on the anode; OH^- loses electrons and releases oxygen,

$$H_2O \rightarrow H^+ + OH^-,$$

$$2OH^- - 2e \rightarrow H_2O + \frac{1}{2}O_2 \uparrow,$$

while zinc ions take up electrons and are gradually deposited on the surface of the cathode,

$$Zn^{2+} + 2e \rightarrow Zn \downarrow .$$

The overall reaction is

$$ZnSO_4 + H_2O \rightarrow Zn \downarrow + H_2SO_4 + \frac{1}{2}O_2 \uparrow .$$

The deposited zinc on the plate is stripped periodically, which is casted as ingot or other zinc products in the casting process. In the electrowinning process, the conversion from zinc ions to solid zinc metal is driven by electric energy. Therefore, the zinc electrowinning process is energy consuming. It accounts for approximately 70% of the overall energy cost of the zinc production process.

1.2.2 Atmospheric direct leaching zinc hydrometallurgy technology

Atmospheric direct leaching zinc hydrometallurgy technology (the ADL process hereafter) is a fully hydrometallurgical process (Fig. 1.8). The main differences between the RLE process and the ADL process include:

(i) Elimination of the roasting process: In the ADL process, the roasting process is eliminated. Zinc sulfide concentrate is grinded to fine particles and fed directly to the leaching process.

(ii) Modified leaching process: In the ADL process, oxygen is blown into the leaching process to oxidize zinc sulfide. The leaching reaction takes place at high temperature and using iron ions as catalyst:

$$ZnS + H_2SO_4 + 0.5O_2 \rightarrow ZnSO_4 + H_2O + S.$$

(iii) Recovery of sulfur: Most of the zinc concentrate is zinc sulfide concentrate. In the RLE process, SO_2 is the resultant of the roasting process, which is then utilized to produce acid. In the ADL process, sulfur is recovered from the leaching residue via sulfur flotation.

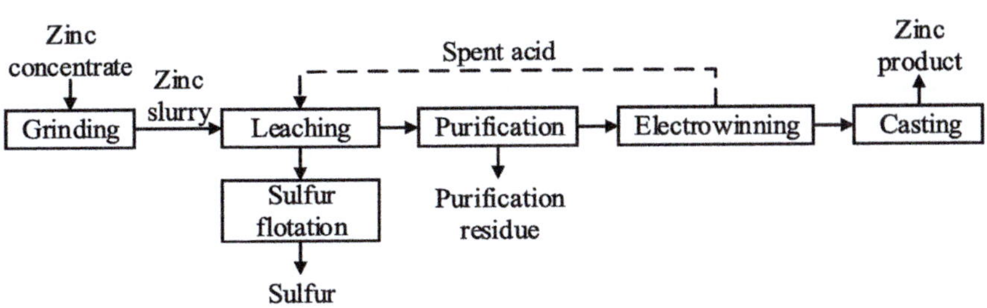

FIGURE 1.8 Atmospheric direct leaching zinc hydrometallurgy technology.

1.3 Solution purification process

The solution purification process is a bottleneck step among all the production stages of zinc hydrometallurgy. As discussed in the previous section, the electrowinning process is energy consuming. The presence of impurity metal ions would degrade the product quality and greatly decrease the efficiency of the electrowinning process [8], increasing the production cost. In some cases, the existence of impurity metal ions even threatens the production safety. Therefore, the impurities remaining after leaching must be purified to certain ranges before electrowinning.

Due to the complexity of zinc concentrate, there exist various types of impurities in the leaching solution. Therefore, the solution purification process is composed of different stages. Each stage removes a particular type of impurity using specific technology. In most zinc hydrometallurgy plants, the solution purification process consists of three stages, i.e., copper removal, cobalt and nickel removal, and cadmium removal. These three stages are shown in Fig. 1.9.

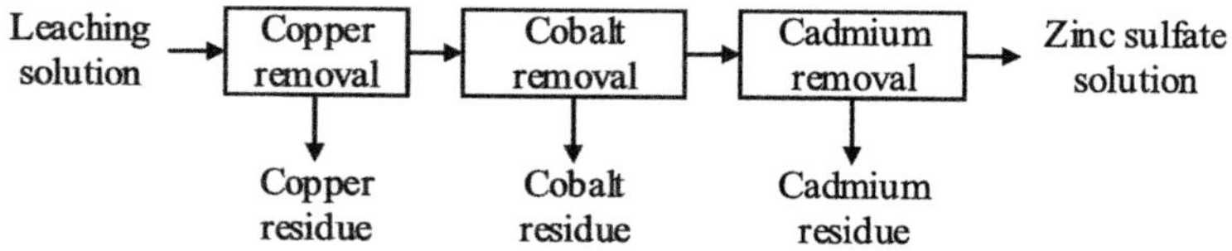

FIGURE 1.9 Solution purification process.

These impurities are removed by adding zinc powder under specific reaction conditions. Via a replacement reaction, the electrons in zinc are transferred to the impurities, which are then cemented on the surface of the zinc powder particles. Among the impurities, copper ions have the highest activity. Therefore, copper removal is the first stage. As copper ions can promote cobalt removal, the aim of the copper removal process is not to remove all the copper ions, but to control the copper ion concentration in a certain range. Cobalt removal is the second stage. Cobalt ions are most detrimental to the electrowinning process, and cobalt removal is the most difficult among the three impurities. Catalysts and high temperature are required to enable cobalt removal. As cobalt and nickel have similar physicochemical properties, they are handled in the same stage and removed using the same technology. The concentration of cobalt ions is higher than that of nickel ions. Therefore, the cobalt and nickel removal process is referred as the cobalt removal process for brevity. The third stage is cadmium removal. Compared with copper and cobalt, cadmium has less influence on the efficiency. It mainly affects the quality of the zinc product. In some plants, cadmium removal is incorporated in the copper or the cobalt removal process. The operational optimization of purification processes includes the determination of zinc powder dosage and the configuration of reaction conditions, such that the production cost is minimized and the technical requirement can be guaranteed. However, due to the comprehensive complexity of the purification process, the operational optimization of the purification process is challenging. Considering the importance of the three stages, this book mainly concerns the modeling, optimization, and control of copper removal and cobalt removal processes.

1.4 Organization and scope of text

This book includes 10 chapters, which are organized in four different parts.

Part I includes Chapters 1 and 2. This part presents the background of the entire book. In Chapter 1, the basic facts, physicochemical properties, and application fields of zinc are introduced. Main zinc hydrometallurgy technologies and a brief outline of the development of zinc hydrometallurgy technology are discussed.

In Chapter 2, the characteristics of the solution purification process and the challenges in the modeling, optimization, and control of the solution purification process are first analyzed. Then, the general modeling and optimal framework of the solution purification process is illustrated.

Part II includes Chapters 3, 4, and 5. This part presents the modeling, optimization, and control of the copper removal process.

In Chapter 3, the flow sheet of the copper removal process is first illustrated. Then, the reaction kinetics of two main chemical reactions in the copper removal process are studied, which form a competitive-consecutive reaction system. Based on the kinetic study, the model structure of the copper removal process is constructed, and the model parameters are identified using an intelligent optimization approach.

In Chapter 4, in order to estimate the zinc powder dosage of each reactor, an additive requirement ratio estimation approach using trend distribution features is proposed.

In Chapter 5, a real-time zinc powder dosage adjustment approach based on the oxidation–reduction potential (ORP) is proposed. The copper removal performance is evaluated by analyzing the ORP, which can be measured online. The concept of controllable domain is proposed, according to which different fuzzy adjustment rules are extracted.

Part III includes Chapters 6, 7, and 8. This part presents the modeling, optimization, and control of the cobalt removal process.

In Chapter 6, an integrated modeling framework is devised. The reaction kinetics of the cobalt removal process which involves multiple electrode reactions is studied in detail. Especially, the relation between the ORP and reaction rate is revealed.

In Chapter 7, an intelligent optimal setting control strategy for the cobalt removal process is proposed. A gradient optimization strategy under normal working conditions and a CBR-based adjustment strategy under abnormal working conditions are studied.

In Chapter 8, two control approaches to determine the zinc powder dosage under different working conditions are proposed. A robust adaptive controller for working conditions with known model parameters but model–plant mismatch is designed. A model-free control based on adaptive dynamic programming is proposed for working conditions with unknown model parameters.

Part IV includes Chapters 9 and 10. This part presents the application, summarizes the entire book, and discusses future research directions in the modeling and optimization of the solution purification process in the context of smart and optimal manufacturing.

References

[1] D.R. Lide, CRC Handbook of Chemistry and Physics, vol. 85, CRC Press, 2004.
[2] J. Meija, T.B. Coplen, M. Berglund, W.A. Brand, P. De Bièvre, M. Gröning, N.E. Holden, J. Irrgeher, R.D. Loss, T. Walczyk, et al., Atomic weights of the elements 2013 (IUPAC technical report), Pure and Applied Chemistry 88 (3) (2016) 265–291.
[3] G. Mei, D. Wang, J. Zhou, H. Wang, Hydrometallurgy of Zinc, Central South University Press, Changsha, 2001.
[4] P. Craddock, The early history of zinc, Endeavour 11 (4) (1987) 183–191.
[5] W. Gui, C. Yang, Intelligent Modeling, Control and Optimization of Complex Nonferrous Metallurgical Production Processes, Science Press, Beijing, 2010.
[6] E. Abkhoshk, E. Jorjani, M. Al-Harahsheh, F. Rashchi, M. Naazeri, Review of the hydrometallurgical processing of non-sulfide zinc ores, Hydrometallurgy 149 (2014) 153–167.
[7] K. Tozawa, T. Nishimura, M. Akahori, M.A. Malaga, Comparison between purification processes for zinc leach solutions with arsenic and antimony trioxides, Hydrometallurgy 30 (1) (1992) 445–461.
[8] O. Bøckman, T. Østvold, Products formed during cobalt cementation on zinc in zinc sulfate electrolytes, Hydrometallurgy 54 (2) (2000) 65–78.
[9] A. Polcaro, S. Palmas, S. Dernini, Kinetics of cobalt cementation on zinc powder, Industrial & Engineering Chemistry Research 34 (1995) 3090–3095.
[10] B. Boyanov, V. Konareva, N. Kolev, Removal of cobalt and nickel from zinc sulphate solutions using activated cementation, Journal of Mining and Metallurgy, Section B: Metallurgy 40 (1) (2004) 41–55.
[11] F. Habashi, A short history of hydrometallurgy, Hydrometallurgy 79 (1–2) (2005) 15–22.
[12] A.D.d. Souza, P.d.S. Pina, V.A. Leão, C.A.d. Silva, P.d.F. Siqueira, The leaching kinetics of a zinc sulphide concentrate in acid ferric sulphate, Hydrometallurgy 89 (1–2) (2007) 72–81.
[13] N. Leclerc, E. Meux, J.M. Lecuire, Hydrometallurgical extraction of zinc from zinc ferrites, Hydrometallurgy 70 (1–3) (2003) 175–183.
[14] S. Fugleberg, A. Jarvinen, E. Yllo, Recent development in solution purification at Outokumpu zinc plant, Kokkola, World Zinc 93 (1) (1993) 241–247.
[15] T. Haakana, M. Lahtinen, H. Takala, M. Ruonala, I. Turunen, Development and modelling of a novel reactor for direct leaching of zinc sulphide concentrates, Chemical Engineering Science 62 (18–20) (2007) 5648–5654.
[16] B.R. Conard, The role of hydrometallurgy in achieving sustainable development, Hydrometallurgy 30 (1–3) (1992) 1–28.
[17] T. Havlík, Hydrometallurgy: Principles and Applications, Elsevier, 2014.

Chapter 2

Modeling and optimal control framework for the solution purification process

Contents

2.1 Problem analysis

The control performance of the solution purification process directly affects the normal functioning and the economical benefits of zinc hydrometallurgy. In theory, the chemistry of the solution purification process is easy to understand, i.e., using zinc powder to replace the impurities under appropriate reaction conditions. However, in practice, the economical and stable operation of this process is difficult to achieve, especially in plants with a low automation level and intricate mineral supplies. Owing to the complex process dynamics and the absence of cheap and reliable online equipment for the detection of metallic ions, operators prefer to handle the solution purification process in a conservative manner, i.e., by using an excessive amount of zinc powder to achieve the required purification performance. Nevertheless, this conservative operation is noneconomical as well as not completely rational. Excessive amounts of zinc powder may cause a local increase in pH and deteriorate the reaction conditions, which could in turn lead to failure in purification. As the design of a reasonable control strategy of an industry process relies on deep understanding of the production technology, control objective, and reaction kinetics, the characteristics of the solution purification process are first analyzed in this section.

Modeling, Optimization, and Control of Zinc Hydrometallurgical Purification Process
https://doi.org/10.1016/B978-0-12-819592-5.00011-9

2.1.1 Challenges in the modeling of the solution purification process

As introduced in Section 1.3, the solution purification process is composed of several impurity removal stages. Each impurity removal stage consists of multiple continuous stirred tank reactors and auxiliary equipment such as storage tank, thickener, and pressure filter. The number of reactors varies in different impurity removal stages. Without loss of generality, impurity removal is conducted in consecutive continuous stirred tank reactors, and zinc powder is used as an additive to replace the target impurity under the assistance of a catalyst, as shown in Fig. 2.1 (B means zinc powder bin, R means reactor, W means weight belt, ST means storage tank, TH means thickener, PF means pressure filter, H means heat exchanger, P means pump).

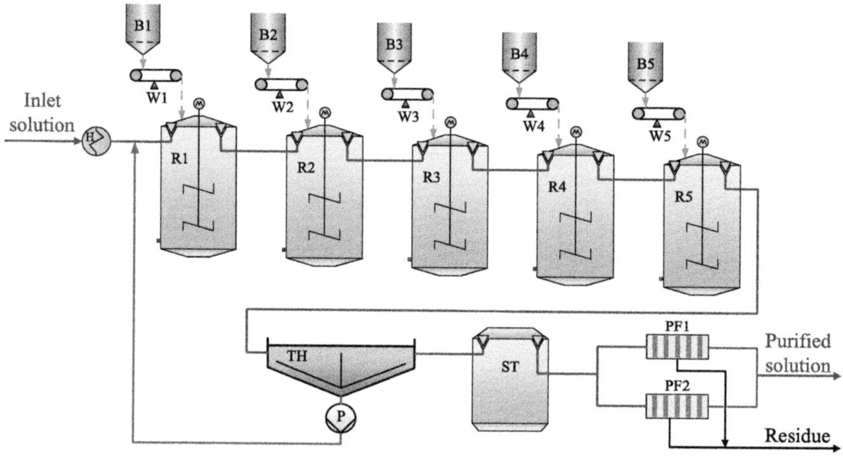

FIGURE 2.1 An impurity removal process.

In each impurity removal stage, the chemical foundation of impurity removal is a replacement reaction between zinc powder and impurity ions, which can be expressed as

$$a\mathrm{Zn} + b\mathrm{Im}^{m+} \rightarrow a\mathrm{Zn}^{n+} + b\mathrm{Im}, \tag{2.1}$$

where Zn is zinc powder, Im is the impurity metal, $a, b, m, n \in \mathrm{N}_{+}$, and $bm = an$ [1]. Zinc powder is used for two reasons:

- Zn is chemically more active than Cu and Co.
- To avoid adding another impurity.

Reaction (2.1) takes place in each reactor. If there is no redissolution of the impurity, then the impurity ion concentration decreases along the cascaded reactors. By replacement, the impurity is cemented on the surface of zinc powder, forming particles of alloys or metal compounds. Therefore, the concentration of impurity ions in the solution is decreased. The particles of alloys or metal compounds are separated from the solution in the thickener. The particles settle down and are recycled to the first reactor. The reasons for the recycling of underflow include:

- These particles contain unreacted zinc powder.
- These particles can act as crystal nuclei to promote impurity removal.

The supernate of the thickener then flows to a storage tank. Then, the solution is further purified using a pressure filter. Finally, the solution flows to next stage to remove another impurity. Although all impurity removal processes share the same chemical foundation, the reaction conditions are different in different stages. For example, in the cobalt removal process, the inlet solution is first heated to approximately 80°C, spent acid is supplied to maintain a suitable pH of the solution, and catalyst is added to enable the occurrence of the cobalt removal reaction. However, in the copper removal process, it is not required to heat the inlet solution.

The operation of an impurity removal stage involves the determination of appropriate reaction conditions and zinc powder dosage of each reactor. Reasonable operation of an impurity process relies on a deep understanding of the process dynamics. However, the dynamics of an impurity removal process has comprehensive complexity, which can be explained from two perspectives:

(i) Inherent complexity of process dynamics: On the one hand, reaction (2.1) is only a simplification or abstraction of the main chemical reactions taking place in an impurity removal process. From a microcosmic perspective, a chemical reaction is essentially a combination of microscopic reactions between molecules. Each microscopic reaction corresponds to a reaction stage of a multistep reaction. On the other hand, due to the heterogeneity of the inlet solution, the chemical reactions conducted in the impurity removal process are not unique. Besides the main reactions, the associated elements/compounds in the inlet solution bring side reactions. These side reactions happen simultaneously and interact with the main reactions in a cooperative or competitive manner, as shown in Fig. 2.2. The process dynamics is affected by the main and side reactions. As there are various types of side reactions, their kinetics and the interaction mechanism with the main reactions are only partially known. Therefore, the intricate interactions between the main and side reactions give rise to the inherent complexity of process dynamics.

(ii) Fluctuation of operating conditions: The impurity removal process can be considered as a unit process of the zinc hydrometallurgy process. As shown in Fig. 2.3, for a unit process, it interacts with other processes through mass/heat transfers, recycling, or reentrances [2]. Therefore, the inlet conditions of an impurity removal process are not constant and can be affected by:

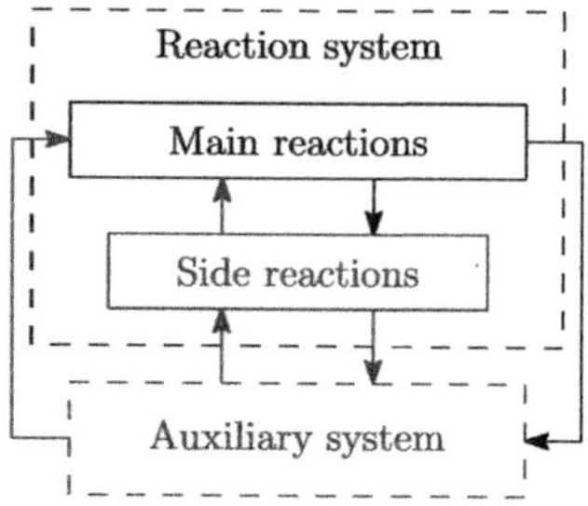

FIGURE 2.2 Interactions among the main reactions, the side reactions, and the auxiliary system.

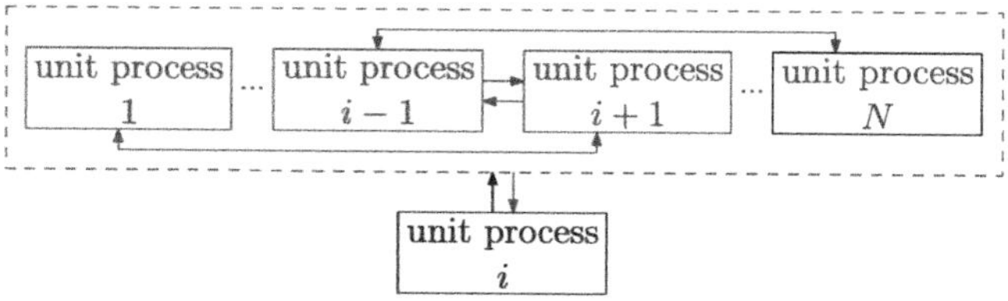

FIGURE 2.3 Interactions among unit processes.

- Changes in the composition of zinc concentrate (if the composition of zinc concentrate changes, then the species in the leaching solution are different).
- Changes in the flow rate of inlet solution.
- Fluctuations of the species concentrations in the inlet solution.
- Plant-wide or local adjustment caused by malfunctions, maintenance, etc.

The types of side reactions change with the composition of inlet solution. In addition, the change in species concentrations of the inlet solution can affect the proportions of main and side reactions. Moreover, the flow rate of inlet solution determines the retention time of the solution in the reactor, which affects the extent of reactions. However, an impurity removal process is composed of not only the main chemical reactors, but also some auxiliary equipments providing necessary reaction conditions, e.g., pH and temperature. For example, in Fig. 2.1, the heat exchanger controls the solution temperature. Chemical reactions exhibit different characteristics under different reaction conditions; variation of reaction conditions can hinder part of the reactions while promoting the rest. Therefore, different combinations of inlet and reaction conditions can form different operating conditions, exhibiting different process dynamics.

To summarize, the dynamics of an impurity removal process is complex and hard to describe. Its dynamics is a synthetical result of the complex interactions among the main reactions, inlet conditions, and reaction conditions, as shown in Fig. 2.4. Therefore, modeling of an impurity removal process requires not only

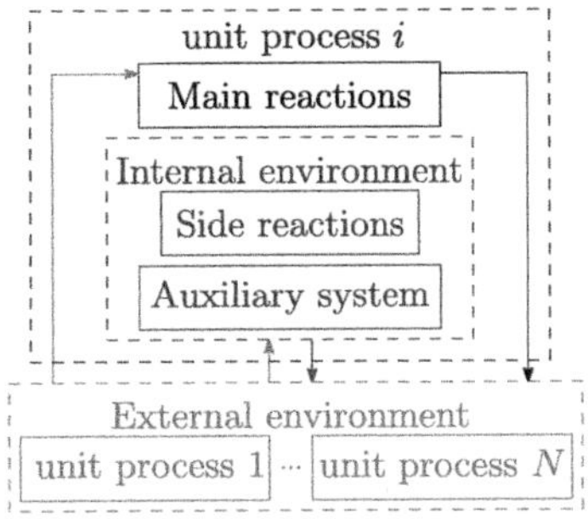

FIGURE 2.4 Interactions among the main reactions, the inlet conditions, and the reaction conditions.

kinetic study of the main reactions, but also design of a framework accounting for the complex interactions and time-varying features.

2.1.2 Challenges in the optimal control of the solution purification process

The economical operation of an impurity removal process involves adjusting the zinc powder dosage (oxidation–reduction potential) and reaction conditions such that the requirement on key performance indicators (KPIs) is met in an economical way, which include:

- The effluent impurity ion concentration of the last reactor is lower than a predefined value.
- The zinc powder consumption is as low as possible.
- The values of outlet impurity ion concentrations of each reactor are as stable as possible.

However, the stable and economical operation of an impurity removal process encounters the following difficulties:

(i) Discrete-delayed measurement of KPIs: Concentrations of the impurities in the leaching solution and the effluent solution of each reactor are key parameters in the determination of zinc powder dosage (oxidation–reduction potential) and reaction conditions. However, online equipment for detecting metallic ions is usually expensive, difficult to maintain in hostile production environments, and not very reliable [3][4]. As a compromise, impurity ion concentrations are usually determined periodically (the sampling period is counted by hours) by a time-consuming artificial chemical assay. Therefore, human operators cannot get access to the real-time information of impurity ion concentrations.

(ii) Multiple operating conditions: In plants with jumbly mineral supplies, the physical and chemical properties of zinc ores vary inevitably. In addition, the operations in the preceding steps of solution purification may not always lead to satisfying results. As a consequence, the inlet conditions, which include the flow rate and types and concentrations of the impurities in the leaching solution, are time-varying [5]. On the other hand, with limited knowledge and information, the operators cannot always adjust the control variables in accordance with the variation of inlet conditions. Thus, as a synthesized outcome of the interlaced variations of inlet conditions, reaction conditions, and manipulated variables, the solution purification process exhibits various kinds of operating conditions. Further, under different operating conditions, the solution purification process has different dynamics and operation targets. Thus, inflexible control strategies are not suitable in this circumstance.

The abovementioned difficulties prevent the operators from making optimal decisions. In order to achieve the desired purification performance, an excessive amount of zinc powder is usually added. However, the relationship between zinc powder dosage and purification performance is not positively correlated. Redundant zinc powder could react with hydrogen ions. This could increase the local pH, resulting in the formation of basic zinc sulfate, which could stick to the surface of zinc powder and hinder impurity removal. This indicates that conservative operation manner is not completely rational. Therefore, a control method that incorporates the online determination of impurity ion concentrations and adapts to the variation of working conditions is required.

2.2 Modeling and optimal control framework

2.2.1 Process modeling based on fusion of reaction kinetics and production data

For complex processes, modeling plays a fundamental role in understanding process dynamics and process control operations. The application scenarios of models include but are not limited to:

- **(i)** Understanding of process dynamics: describing the physical and chemical phenomena of a process under both steady and unsteady states, and representing the relationship between measured process variables and technical indices.
- **(ii)** Process monitoring: soft-sensing, fault detection and diagnosis (FDD), control performance evaluation, etc.
- **(iii)** Controller design and evaluation: developing a simulation environment to test control strategies.
- **(iv)** Operational optimization: determination of appropriate operation points under different working conditions.

Modeling of a process is essentially deriving a set of mathematical formulations to achieve the above functions. There exist various types of models. Ljung used a palette of gray shades from white to black to categorize models [6]:

- White model: first-principle models described by differential/algebraic equations (DAEs).
- Off-white model: white models with unknown or uncertain parameters described by, e.g., state space models.
- Smoke-gray model: semi-physical models.
- Steel-gray model: models linearized around an operating point.
- Slate-gray model: hybrid models, block-oriented models.
- Black model: neural network, support vector machines (SVMs).

Essentially, all these models approximate the process dynamics by learning from data, information, or knowledge related to the process, e.g., operation data, reaction kinetics, first principles.

The model performance relies on the selection of an appropriate model structure and effective utilization of operation data, reaction kinetics, and first principles. These above models use different sources of data, information or knowledge, and have their pros and cons. As the dynamics of the impurity removal process are highly complex, it is difficult for a single model to describe all the facets of process dynamics. Moreover, in the era of smart and optimal manufacturing, a descriptive system should be able to not only act as a model describing the process dynamics, but also:

- Act as a platform supporting digitalization and visualization.
- Systematically support process control applications.

Therefore, the design of a descriptive system and a corresponding modeling framework is required. Such a descriptive system should be capable of supporting smart and optimal manufacturing, while the modeling framework should be able to incorporate different types of model that increase the utilization rate of data, information, and knowledge related to the process.

2.2.1.1 Definition of a comprehensive state space descriptive system

If we only consider the main reactions of an impurity removal process, its nominal dynamics can be derived by studying the reaction kinetics. The resulted process model is a state space model with the states being the technical indices:

$$\dot{\mathbf{x}}_{\mathrm{O}} = \mathbf{f}(\mathbf{x}_{\mathrm{O}}, \Theta_{\mathrm{F}}), \tag{2.2}$$

where $\mathbf{x}_{\mathrm{O}}$ denotes the technical indices, $\mathbf{f}(\cdot)$ is the nominal kinetic model, and Θ_{F} is the model parameters.

As shown in Fig. 2.5, the practical process dynamics of impurity removal are an outcome of the interactions between main reactions, reaction conditions, and inlet conditions. If taking the complex interactions into account, the dynamics of the technical indices process can be expressed by

$$\dot{\mathbf{x}}_{\mathrm{O}} = \mathbf{g}(\mathbf{x}_{\mathrm{O}}, \mathbf{x}_{\mathrm{I}}, \mathbf{x}_{\mathrm{R}}), \tag{2.3}$$

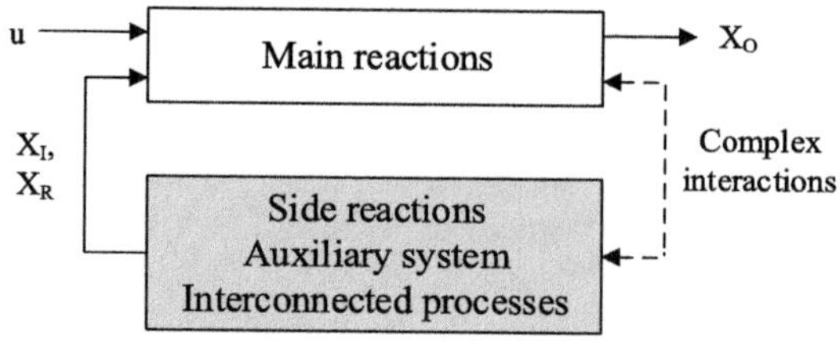

FIGURE 2.5 Practical process dynamics.

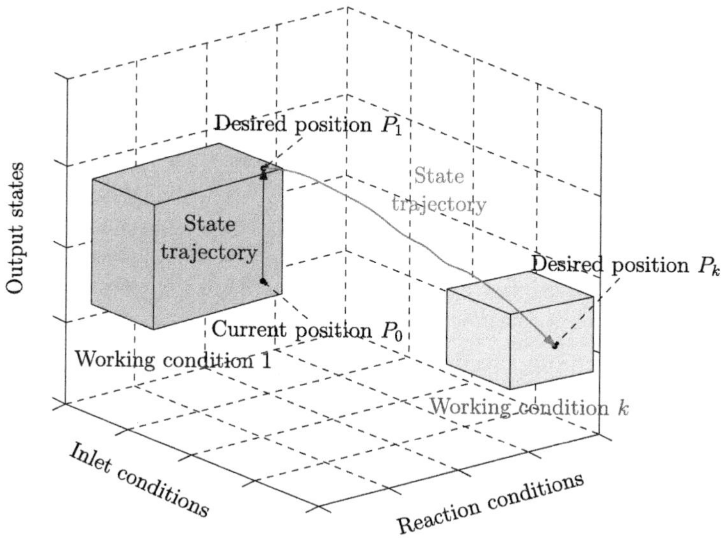

FIGURE 2.6 Comprehensive state space (P_0, P_1, and P_k are different operating points).

where $\mathbf{x}_I$ and $\mathbf{x}_R$ denote the inlet conditions and the reaction conditions, respectively, and $\mathbf{g}(\cdot)$ is a function representing the relation between $\dot{\mathbf{x}}_O$ and $\mathbf{x}_O$, $\mathbf{x}_I$, and $\mathbf{x}_R$. Therefore, by augmenting $\mathbf{x}_O$ with $\mathbf{x}_I$ and $\mathbf{x}_R$, the original state space containing only the technical indices can be extended to a "comprehensive state space" (CSS) (Fig. 2.6) which explicitly includes more influencing factors of the process dynamics. The comprehensive state space descriptive system is defined as follows [7].

Comprehensive state space: Comprehensive state space is a three-dimensional vector space with each dimension being the codes of:

- Output states ($\mathbf{x}_O$): the controlled technical indices of an impurity removal process, i.e., the outlet impurity ion concentrations of each reactor.
- Inlet conditions ($\mathbf{x}_I$): the flow rate and composition/species concentrations of the inlet solution.
- Reaction conditions ($\mathbf{x}_R$): the conditions under which the main and side reactions take place, including temperature, pH, etc.

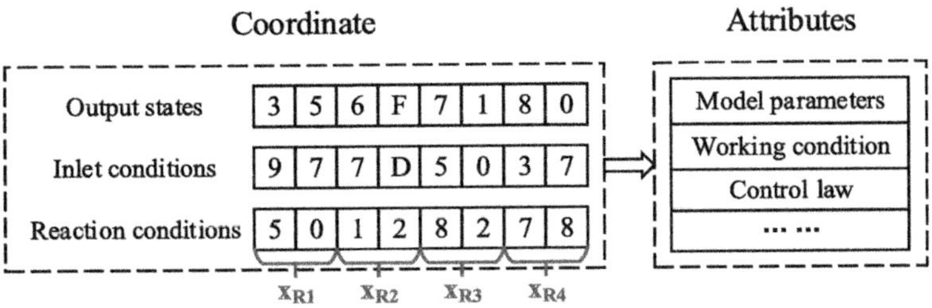

FIGURE 2.7 Coordinate and attributes of a point in the comprehensive state space.

The three vectors $\mathbf{x}_O$, $\mathbf{x}_I$, and $\mathbf{x}_R$ are high-dimensional. In order to construct a three-dimensional CSS, each of the three vectors has to be coded into a one-dimensional variable [8]. This can be achieved if:

- Each axis/dimension of the CSS is expressed as a code which is a combination of hexadecimal numbers.
- For each axis/dimension of the CSS, different parts of the code are assigned to different elements of the original high-dimensional vectors.

As shown in Fig. 2.7, the axis of reaction conditions $\mathbf{x}_R$ is composed of four hexadecimal numbers, i.e., $\mathbf{x}_{R1}$, $\mathbf{x}_{R2}$, $\mathbf{x}_{R3}$, and $\mathbf{x}_{R4}$, representing temperature, flow rate of spent acid (times 10), flow rate of arsenic trioxide, and solid content in the thickener, respectively. All these physical quantities have two digits, e.g., take the value of 80, 1.8, 130, and 120 respectively.

The CSS description system covers more influencing factors than the traditional state space composed solely of output states, which reduces the uncertainty in modeling. In addition, as the three axes of the CSS are combinations of hexadecimal numbers, the CSS descriptive system is a digitalized descriptive system. As shown in Figs. 2.6 and 2.7, at each time point, $\mathbf{x}_O$, $\mathbf{x}_I$, and $\mathbf{x}_R$ have their own values and correspond to the coordinate of a point in the CSS. Moreover, one can assign different attributes to the point in the CSS, including model structure, model parameters, working condition, control law, etc. (Fig. 2.7). These attributes can describe the dynamics of a point in the CSS and provide extra information supporting process control applications. Therefore, for each point, its coordinate and associated attributes form a comprehensive description [7].

The CSS descriptive system is a feasible option to support modeling and smart and intelligent manufacturing of complex industry processes, which can be explained from the following aspects [7].

(i) Comprehensive description ability: The practical dynamics of a process is a synthesized outcome of the intricate interactions among main reactions, inlet conditions, and reaction conditions. In the CSS, the controlled technical indices $\mathbf{x}_O$, inlet conditions $\mathbf{x}_I$, and reaction conditions $\mathbf{x}_R$ are incorporated in a three-dimensional coordinate system. Therefore, the CSS descriptive system comprehensively covers the main influencing factors of the process dynamics. The variation of $\mathbf{x}_O$ against time is formulated as a function of $\mathbf{x}_O$, $\mathbf{x}_I$, and $\mathbf{x}_R$. Therefore, considering the technical indices $\mathbf{x}_O(t_1) = \mathbf{x}_O(t_2)$ for two time intervals $[t_1, t_1 + \Delta t]$ and $[t_2, t_2 + \Delta t]$, if the initial values and variations of $\mathbf{x}_I$ and $\mathbf{x}_R$ are the same over these two time intervals, then $\mathbf{x}_O(t_1 + \Delta t) = \mathbf{x}_O(t_2 + \Delta t)$.

(ii) Digitalized nature: The three axes of the CSS are combinations of hexadecimal numbers. The attributes of a point in the CSS are either data, information, or knowledge which can be obtained and stored in an attribute database. So the CSS descriptive system is digitalized in nature. The "coordinate-attributes" information of each point in the CSS is a digitalized description of the point. The operator can access the value of model parameters, the type of working condition, and the configuration of the optimal controller of the process by obtaining the coordinate and searching this information in the attribute database. The digitalized description contains rich information about the dynamics of the process and can then be used to develop a digital twin of the physical process.

(iii) Capable of visualization: In a sufficiently small subspace of the CSS, the model parameters of the points inside are close enough. Therefore, the CSS can be divided into subspaces in which the process dynamics can be regarded as constant. On the other hand, in different subspaces, $\mathbf{x}_O$, $\mathbf{x}_I$, and $\mathbf{x}_R$ are in different value ranges, and the process exhibits different dynamics. So these subspaces often correspond to different working conditions. By subdividing the CSS into subspaces, the boundary of working conditions can be visualized. Therefore, one can observe the transition of working conditions along the movement trajectory of the process.

2.2.1.2 *Typical modeling approaches*

There exist various types of models, depending on the formulation and methods of the model, e.g., I/O model, finite impulse response (FIR), Auto-Regressive with eXogenous inputs (ARX), step response model, state space model, Auto-Regressive Moving Average with eXogenous inputs (ARMAX), Box–Jenkins model (more general than ARMAX), Hammerstein model, Wiener model, Nonlinear Auto-Regressive with eXogenous inputs (NARX), NARMAX, Controlled Auto-Regressive Integrated Moving Average (CARIMA), Volterra-series model, Carleman approximation model, Wiener–Hammerstein model, Nonlinear Auto-Regressive with eXogenous Inputs, linear models (partial least squares [PLS]), Auto-Regressive Integrated Moving Average model (ARIMA), and U-model. Among these models, the state space-based first-principle model (SS-FPM) and machine learning-based input/output model (ML-IOM) are two typical models [7]. Before presenting the hybrid modeling framework, these two modeling approaches are briefly reviewed and compared in this subsection.

State space-based first-principle modeling

First-principle models are often derived based on conservation laws and the underling reaction mechanism of a process. The conservation laws include conservation of mass, conservation of components, conservation of energy, etc. [9]. The reaction mechanism can be described by reaction kinetics, thermodynamics, etc. The conservation laws describe the accumulation rate of mass/component/energy inside a reactor/process. For example, the conservation law of a component M inside a reactor can be formulated as follows:

$$\frac{dM}{dt} = M_{in} - M_{out} - M_r, \tag{2.4}$$

where M is the mass/concentration of the component inside the reactor, M_{in} is the rate with which the component flows into the reactor, M_{out} is the outflow rate of the component, and M_r indicates the consumption rate of the component. Considering a reactor, a component is consumed by acting as reactant in certain chemical reactions; its consumption rate is determined by the mechanism of the reactions in which it is involved. Therefore, M_r can be derived by studying the underlying reaction mechanism. This type of modeling approach is also known as physical modeling, which derives model equations from physical knowledge [10]. It is noted that Eq. (2.4) is a differential equation expressing the variation of M against time, which can be considered as a state space model.

The concept of "state" was proposed in R.E. Kalman's seminal work "*A New Approach to Linear Filtering and Prediction Problems*," published in 1960 [11]. The aim of this work is to solve an optimal estimation problem or the Wiener problem. The optimal estimation problem is a duality of the noise-free optimal regulator problem, which belongs to the realm of probability theory and statistics [11]. In [11], state is defined as follows:

- *State is the least amount of data one has to know about the past behavior of the system in order to predict its future behavior. The dynamics is then described in terms of state transitions, i.e., one must specify how one state is transformed into another as time passes.*

Therefore, Eq. (2.4), M, which is the mass/concentration of the component inside the reactor, can be regarded as a state.

A state space model is a set of differential/difference algebraic equations (DAEs) of states, inputs, and outputs. The inputs are manipulated variables. The outputs are measured economic or technical indices or functions of states and inputs. As shown below, linear and nonlinear state space models are formulated as

$$\begin{aligned} \dot{x} &= Ax + Bu, \\ y &= Cx + Du, \end{aligned} \tag{2.5}$$

$$\begin{aligned} \dot{x} &= f(x, u), \\ y &= g(x, u), \end{aligned} \tag{2.6}$$

where A, B, C, and D are dynamics matrix, control matrix, sensor matrix, and direct matrix, respectively, and $f(x, u)$ and $g(x, u)$ are smooth nonlinear functions. In this study, as the states and outputs represent the same physical quantities, the output equation can then be eliminated. In addition, as the state space model is built based on conservation laws and the reaction mechanism, the model parameters have their physical meaning [12].

Consider that an impurity removal process contains $N (N \in \mathbb{N}^+)$ reactors and assume that the fluid in each reactor is perfectly mixed, i.e., the concentrations, temperature, and reaction rate are the same everywhere in the entire reactor. Then, according to the mass balance principle, the nominal kinetic model of the impurity removal process is

$$\frac{dc_i}{dt} = \frac{F_{i-1}}{V} c_{i-1} - \frac{F_i}{V} c_i - k_i A_i c_i, \tag{2.7}$$

where $i = 1, 2, \cdots, N$, V is the volume of the reactor, c_{i-1} and c_i are the influent and effluent impurity ion concentrations of the ith reactor, respectively, k_i and A_i are the reaction rate constant and reaction surface area in unit volume of the ith reactor, respectively, and F_{i-1} and F_i are the influent and effluent flow rates of the ith reactor, respectively. In particular, $F_0 = F_{in} + F_u$, where F_{in} is the flow rate of the input leaching solution from the previous stage and F_u is the flow rate of the recycled underflow. The unknown parameters $\theta = [A_{\mathrm{F}}, E_{\mathrm{e}}, \gamma, e_{\mathrm{eq}}, A]$ in the nominal model can be identified from the historical production data.

The advantages of SS-FPM include the following:

- **(i)** It is constructed based on the conservation laws and reaction mechanism, so it represents the underlying physical or chemical laws of the process.
- **(ii)** It explicitly formulates the relationship between inputs and states, so it is widely used in controller design.

Its disadvantages include the following:

- **(i)** The key of SS-FPM lies in studying the reaction mechanism, which is time consuming and requires lots of domain knowledge.
- **(ii)** The laws or principles used in deriving SS-FPM are based on certain assumptions or obtained under specific experimental conditions. However, in practice, the reaction and inlet conditions are time-varying. The essential principles in the SS-FPM may still hold, while the model parameters vary with the reaction and inlet conditions.
- **(iii)** The SS-FPM may not be able to explicitly include all the influencing factors of the process dynamics.

Machine learning-based input/output modeling

In machine learning-based input/output models (ML-IOMs), the inputs are the influencing factors of process dynamics, while the outputs are the concerned technical indices. So the ML-IOM represents the statistical correlations between inputs and outputs. It first uses machine learning approaches (unsupervised learning, supervised learning, semisupervised learning, reinforcement learning) to extract latent variables or features from the original high-dimensional process variables [13]. Then, it reconstructs the technical indices from those latent variables. These machine learning approaches have their roots in probability theory and statistics.

ML-IOM involves the use of machine learning approaches which project the real process to a data space or latent variable space and construct statistical models for correlation analysis, prediction, soft sensing, process monitoring, pattern recognition, and FDD [14]. These approaches can be categorized as:

(i) Unsupervised learning.
(ii) Supervised learning.
(iii) Semisupervised learning.
(iv) Reinforcement learning.

The main actions involved in ML-IOM approaches include:

(i) Regression.
(ii) Classification.
(iii) Clustering.
(iv) Coordinate transformation.
(v) Statistical properties analysis.

These methods make use of a large volume of production data that contain useful information and knowledge about the process dynamics and working conditions. In addition, by projection, the evolution of operating conditions can be visualized in the data space. However, the performance of data-driven modeling relies on the quality and operation ranges covered by the training data.

Comparison between SS-FPM and ML-IOM

As shown in Fig. 2.8, the two modeling methods SS-FPM and ML-IOM use different sources of data, information, and knowledge, and have their pros and cons:

(i) The SS-FPM utilizes inherent physical or chemical laws such as the underlying physicochemical laws of a process. However, not all the factors that influence the process are captured in the model.
(ii) The ML-IOM utilizes routinely collected production data and machine learning algorithms. It can approximate the relationship between process inputs and outputs with very high accuracy. However, it does not capture the internal dynamics of a system.

However, both methods are informed by probability theory and statistics, e.g., conservation laws are derived from experimental data, and first-principle models often contain empirical relationships. The combination of these two methods could increase the utilization rate of data, information, and knowledge, thus leading to the discovery of more facets of the process dynamics.

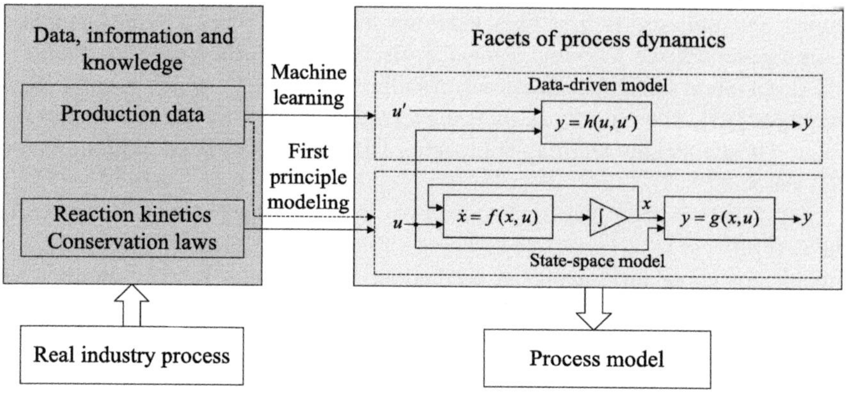

FIGURE 2.8 Differences and connections between state space and data-driven modeling.

2.2.1.3 Hybrid first-principle/machine learning modeling frameworks

As discussed above, the fundamentals of modeling include the selection of a model structure and utilization of data, information, and knowledge associated with process dynamics. Different modeling approaches utilize different sources of data/information/knowledge, which reveal the process dynamics from different aspects. Therefore, finding an appropriate model structure which can accommodate different sources of data/information/knowledge in a framework is important for increasing the accuracy and performance stability of a model. In this section, several typical hybrid model frameworks are introduced.

Naive integration of a kinetic model and a data-driven compensation model

The kinetic model forces on the main reactions. It can describe the process dynamics qualitatively. However, impurity removal is a complex multiphase reaction influenced by numerous factors. The complex influencing mechanism of various factors cannot be completely described by the kinetic model. In addition, the theorem and principle adopted in deducing the kinetic model are based on assumptions and obtained under specific conditions. The practical process dynamics are not exactly the same with each kinetic model, i.e., there exists deviation between kinetic models and process dynamics.

Application of a distributed control system (DCS) and a fieldbus control system (FCS) in metallurgy plants enables the collection of daily operation data which contains abundant information about the process dynamics. Data-driven compensation uses regression methods to learn from the data, i.e., train a model using the data which could reconstruct the relationship between important process variables and KPIs of interest.

In order to combine the kinetic model and the data-driven compensation model, the system dynamics described by (2.3) can be decomposed as a combination of a nominal term and a deviation term:

$$\dot{\mathbf{x}}_{\mathrm{O}} = \mathbf{f}(\mathbf{x}_{\mathrm{O}}, \Theta, \mathbf{u}) + \boldsymbol{\epsilon}(\mathbf{x}_{\mathrm{O}}, \mathbf{x}_{\mathrm{I}}, \mathbf{x}_{\mathrm{R}}), \tag{2.8}$$

where $\mathbf{f}(\cdot)$ is the nominal term or the kinetic model containing the main reactions (white model) and Θ represents the model parameters, which can be identified using historical operation data. This formulation keeps the nominal dynamics of the main reactions in a state space form, and can account for the intricate interactions by introducing $\epsilon(\cdot)$, which then needs to be determined using data analytic methods (gray or black model), as shown in Fig. 2.9. This model approach utilizes both mechanism knowledge and process measurements.

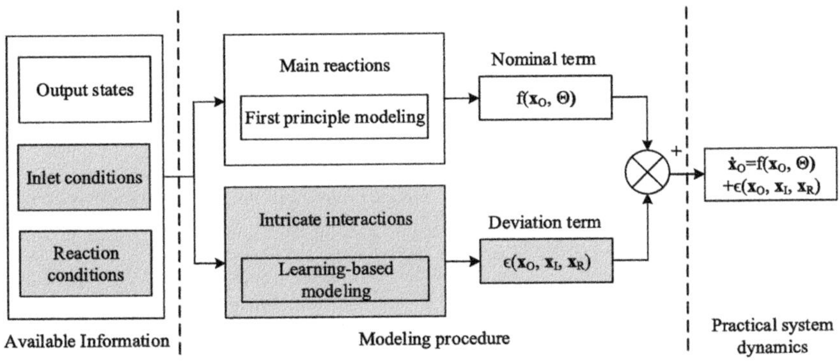

FIGURE 2.9 Naive integration of a kinetic model and a data-driven compensation model.

Considering each operating condition, the process dynamics can be formulated as follows:

$$\mathbf{x}_{\mathrm{O}}(t_f) = \mathbf{x}_{\mathrm{O}}^{\mathrm{nominal}}(t_f) + \mathbf{x}_{\mathrm{O}}^{\mathrm{deviation}}(t_f), \tag{2.9}$$

where

$$\mathbf{x}_{\mathrm{O}}^{\mathrm{nominal}}(t_f) = \mathbf{x}_{\mathrm{O}}^{\mathrm{nominal}}(t_0) + \int_{t_0}^{t_f} \mathbf{f}(\mathbf{x}_{\mathrm{O}}, \Theta)dt, \tag{2.10}$$

$$\mathbf{x}_{\mathrm{O}}^{\mathrm{deviation}}(t_f) = \mathbf{x}_{\mathrm{O}}^{\mathrm{deviation}}(t_0) + \int_{t_0}^{t_f} \boldsymbol{\epsilon}(\mathbf{x}_{\mathrm{O}}, \mathbf{x}_{\mathrm{I}}, \mathbf{x}_{\mathrm{R}})dt, \tag{2.11}$$

$\mathbf{x}_{\mathrm{O}}^{\mathrm{nominal}}(t_f)$ is the nominal term derived from the FPM of the main reactions [15], [16], and $\mathbf{x}_{\mathrm{O}}^{\mathrm{deviation}}(t_f)$ is the deviation term with the output states $\mathbf{x}_{\mathrm{O}}$, inlet conditions $\mathbf{x}_{\mathrm{I}}$, and reaction conditions $\mathbf{x}_{\mathrm{R}}$ as its inputs. Thus, the value of $\mathbf{x}_{\mathrm{O}}$ can be expressed as

$$\mathbf{x}_{\mathrm{O}}(t_f) = \mathbf{x}_{\mathrm{O}}(t_0) + \int_{t_0}^{t_f} [\mathbf{f}(\mathbf{x}_{\mathrm{O}}, \Theta) + \boldsymbol{\epsilon}(\mathbf{x}_{\mathrm{O}}, \mathbf{x}_{\mathrm{I}}, \mathbf{x}_{\mathrm{R}})]dt, \tag{2.12}$$

or, more practically, in a nonlinear discrete form,

$$\mathbf{x}_O(k+1) = \mathbf{x}_O(k) + \mathbf{f}_k(\mathbf{x}_O(k), \Theta(k)) + \epsilon_k(\mathbf{x}_O(k), \mathbf{x}_I(k), \mathbf{x}_R(k)), \qquad (2.13)$$

where k indicates the kth sampling instant and $\mathbf{f}_k(\mathbf{x}_O(k), \Theta(k))$ and $\epsilon_k(\mathbf{x}_O(k), \mathbf{x}_I(k), \mathbf{x}_R(k))$ are discretized counterparts of $\int_{t_k}^{t_{k+1}} \mathbf{f}(\mathbf{x}_O, \Theta)dt$ and $\int_{t_k}^{t_{k+1}} \epsilon(\mathbf{x}_O, \mathbf{x}_I, \mathbf{x}_R)dt$, respectively.

In Eq. (2.13), $\epsilon_k(\mathbf{x}_O(k), \mathbf{x}_I(k), \mathbf{x}_R(k))$ is the data-driven compensation term which usually takes a nonlinear regression form,

$$\epsilon_k = \mathbf{U}(\mathbf{z}(k)) = [\mathrm{U}_1(\mathbf{z}(k)) \quad \mathrm{U}_2(\mathbf{z}(k)) \cdots \mathrm{U}_N(\mathbf{z}(k))]^{\mathrm{T}}, \qquad (2.14)$$

where $\mathbf{z}$ is the vector of model inputs; $\mathbf{z}$ can be raw measurements of $\mathbf{x}_O(k)$, $\mathbf{x}_I(k)$, and $\mathbf{x}_R(k)$, or low-dimensional deep process features extracted from the raw measurements; and $\mathbf{U}(\cdot)$ is the regression function set of $\mathbf{z}$.

Integration of a subkinetic model and a subdata-driven compensation model

As discussed in Section 2.1.1, an impurity removal process has various working conditions. Owing to the diversity and variability of the working conditions, the nominal kinetic model and the data-driven compensation model with fixed parameters may possibly generate a large bias in KPI prediction.

Impurity removal exhibits various working conditions. Under different working conditions, model parameters take different values. Therefore, industrial data is classified into several working condition classes, then kinetic model parameters are identified, and the data-driven compensation model of each working condition is trained. If there exist sufficient amounts of data samples for each working condition, then this approach can be applied to improve the modeling accuracy, as shown in Fig. 2.10. This modeling approach utilizes not only the mechanism knowledge and process measurements, but also working condition information.

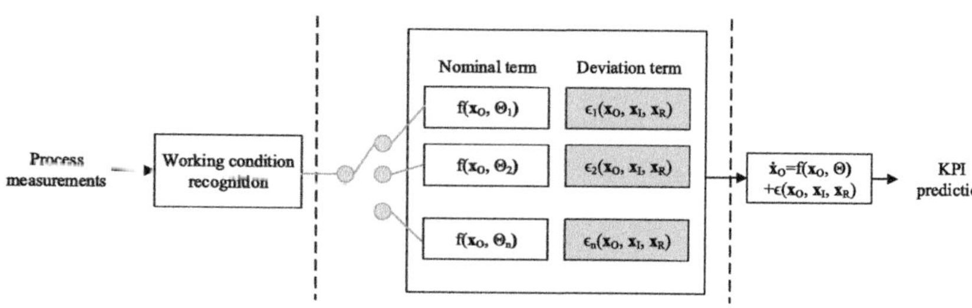

FIGURE 2.10 Integration of a subkinetic model and a subdata-driven compensation model.

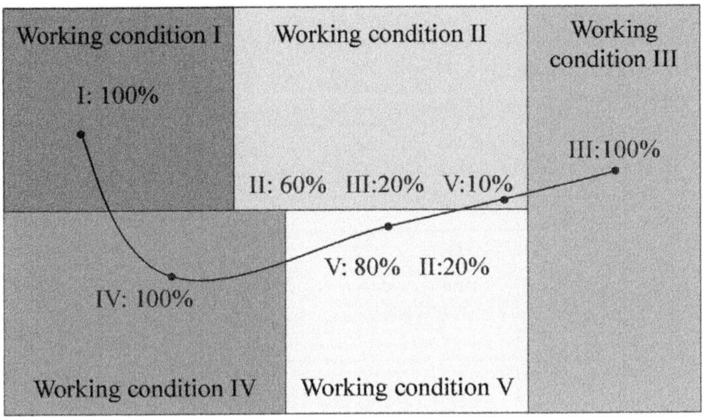

FIGURE 2.11 Evolution trajectory of an impurity removal process.

Weighted hybrid kinetic model and data-driven compensation model with time-varying weights

As the process evolves in the CSS, the model parameters are subject to the inlet conditions, reaction conditions, and output states. However, these parameters are identified for each operating condition, not each point in the CSS. Considering the time-varying characteristic of the production environment, there exists a matching degree between the current process and different working conditions (Fig. 2.11), especially when the process is located at the junction of different working conditions. A weighted hybrid modeling approach with time-varying weights has been proposed, which can be expressed as

$$\mathbf{x}_O(k+1) = \mathbf{x}_O(k) + \sum_{i=1}^{N_p} P(i, \mathbf{z})[\mathbf{f}_k^{(i)}(\mathbf{x}_O(k), \Theta(k)) + \mathbf{U}^{(i)}(\mathbf{z}(k))], \qquad (2.15)$$

where $\mathbf{f}_k^{(i)}(\mathbf{x}_O(k), \Theta(k))$ and $\mathbf{U}^{(i)}(\mathbf{z}(k))$ indicate the corresponding terms associated with the kinetic model and the compensation model under operating condition i and $P(i, \mathbf{z})$ indicates the matching degree between the current process and working condition i. The value of $P(i, \mathbf{z})$ varies with the location of the current process in the CSS. Compared with the previous modeling approach, the proposed CSS-based modeling approach not only considers the multimodality, but also updates the weights of each submodel to improve the stability of model performance.

Comprehensive hybrid modeling framework

In the industrial field, reasons like emerging of new working conditions and sensor drift may deteriorate the model performance. Therefore, it is necessary to update the kinetic model or data-driven compensation model online to improve

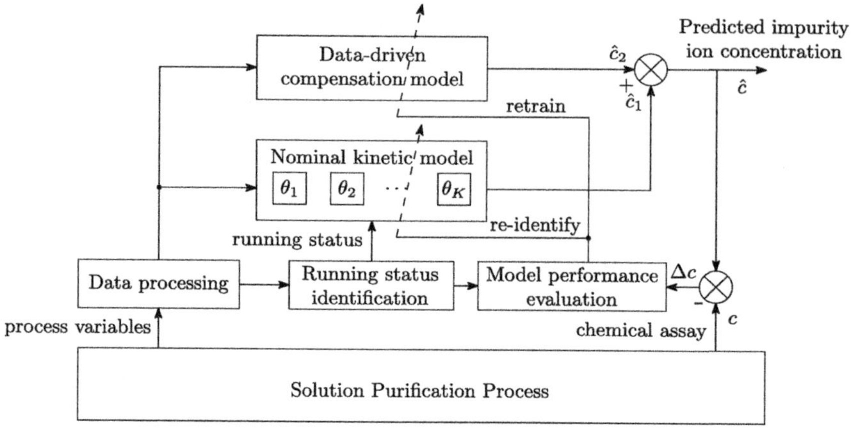

FIGURE 2.12 Multiple time-scales model compensation.

the adaptability. Model updating is reidentifying the kinetic model parameters or retain the data-driven compensation model periodically or when the model error is larger than a predefined limit. Model updating can be incorporated into the weighted hybrid modeling approach to construct a comprehensive hybrid modeling framework, as shown in Fig. 2.12.

The framework contains two stages: an offline stage and an online stage. The preparation work is done in the offline stage, which mainly utilizes the historical data covering a wide operation range. The kinetic model parameters of each working condition $\{\theta_1, \theta_2, \cdots, \theta_n\}$ ($n \in \mathbb{N}^+$ is the number of working conditions) are identified. The data-driven compensation models of each working condition are trained using historical data. In the online stage, the real-time process measurements, which reveal the current location of an impurity removal process in its CSS, are fed as input of the weight hybrid model. The performance of the weighted hybrid model is evaluated continuously based on the mismatch between its output and the actual KPI, or more specifically:

(i) If the degree of mismatch is less than ε_{low}, then no model updating is required.

(ii) If the degree of mismatch is between ε_{low} and ε_{median}, then the data-driven compensation models are retrained.

(iii) If the degree of mismatch is larger than ε_{median}, then the parameters of the kinetic models are reidentified.

2.2.2 Cooperative optimization and control of cascaded metallurgical reactors

The optimal control of an impurity removal process is a constraint optimization problem. The objective of this problem is to find the best combination of zinc powder dosages of each reactor

$$\min_{\mathbf{u}} \quad J = \sum_{i=1}^{N} u_i, \tag{2.16}$$

which minimizes the overall zinc powder consumption and guarantees that the effluent impurity ion concentration of the last reactor is within a required range,

$$c_{\text{out,min}} \leqslant c_{\text{out}} \leqslant c_{\text{out,max}}, \tag{2.17}$$

where $\mathbf{u} = [u_1, u_2, \cdots, u_N]$, u_i is the zinc powder dosage of the ith reactor, c_{out} is the effluent impurity ion concentration of the last reactor, and $c_{\text{out,min}}$ and $c_{\text{out,max}}$ are the lower and upper limit of c_{out}. As an impurity removal is composed of cascaded reactors, besides the above basic operation objective, the outlet impurity ion concentration of each reactor should be kept in a reasonable range to guarantee the stability of each reactor. Therefore, for the ith reactor,

$$c^i_{\text{out,min}} \leqslant c^i_{\text{out}} \leqslant c^i_{\text{out,max}}. \tag{2.18}$$

This is reasonable as large and frequent variations of KPIs would cause the instability of a system.

The optimal control of a plant is achieved by decomposing the optimal control problem into multiple function levels ranging from production planning to single-loop control. As an impurity removal process is composed of cascaded reactors, its optimal control problem (2.16) can be decomposed into the economical optimization of the entire process and the closed-loop optimal control of each reactor.

Generally, the design of an economical optimization strategy depends on a deep understanding of characteristics of a process. The optimization strategy should be compatible with the process to be optimized. For an impurity removal process, its overall impurity removal task is allocated to each reactor with different impurity removal abilities. However, due to the fluctuations of inlet and reaction conditions, only keeping the KPIs of each reactor constant cannot guarantee the optimal operation of the entire process. Therefore, the KPIs of each reactor should be optimized periodically according to the current inlet and reaction conditions. As shown in Fig. 2.13, the economical optimization of the entire impurity removal process is essentially the cooperative optimization of cascaded reactors, which is equivalent to finding the best KPIs of each reactor. The problem can be expressed as

$$\min_{\mathbf{c}} \quad J = f(\mathbf{c}), \tag{2.19}$$

where $\mathbf{c} = [c_1, c_2, \cdots, c_N]$ is the outlet impurity ion concentration of each reactor, or the KPIs of each reactor. This optimization problem (2.19) is solved periodically to provide set-points of KPIs. Please note that the optimization

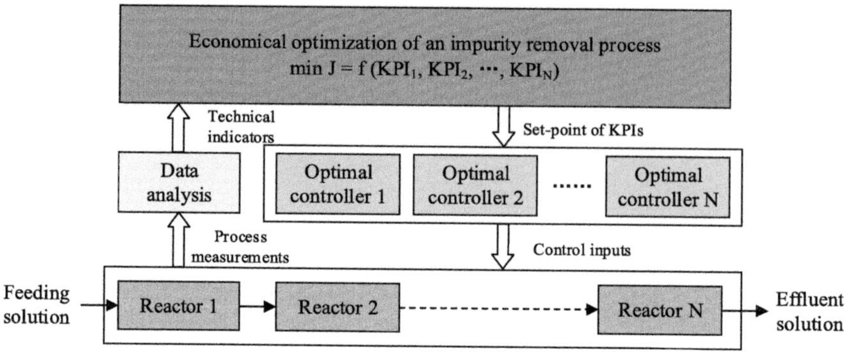

FIGURE 2.13 Optimization and control of an impurity removal process.

problem (2.19) is still subject to the above constraints. The transformation is achieved by designing technical indicators which can be used to reformulate the economical optimization problem to the optimization problem of KPIs of each reactor. The economical optimization problem is introduced in detail in Chapter 4 and Chapter 7.

Then, the closed-loop optimal control of each reactor is to determine the value of control input which forces the practical KPIs of each reactor to follow their setting values. Then, for each reactor, its optimal control problem is setting value

$$\min_{u} \quad J_{\text{Reactor}} = \int_{t_0}^{t_{\text{final}}} ||c(u) - c^*||^2 + M(u)dt, \tag{2.20}$$
$$\text{s.t.} \quad c_{\min} \leqslant c \leqslant c_{\max},$$
$$u_{\min} \leqslant u \leqslant u_{\max},$$

where $c(u)$ is the practical value of outlet impurity ion concentration, c^* is the set-point of c, $[c_{\min}, c_{\max}]$, $[u_{\min}, u_{\max}]$ are the constraints of KPI and control input, $[t_0, t_{\text{final}}]$ is the control interval of interest, and $M(u)$ is the function measuring the control cost.

The problem (2.20) is essentially an optimal control problem. As an impurity removal process has multiple working conditions, this problem takes different forms under different working conditions. When the process is under working conditions with clearly known model parameters, then the problem (2.20) can be solved using model-based optimal control approaches, e.g., model predictive control. When the process is under working conditions with model uncertainties, then the problem (2.20) is transformed to designing a stabilizing controller, e.g., robust adaptive control, or intelligent operational optimization approaches. When the process is under working conditions with unknown model parameters, then the problem (2.20) can be solved using reinforcement learning-based optimal control approaches, e.g., adaptive dynamic programming. In Chapters 5 and 8, the authors summarize their previous results on solving the control problem of the copper removal process and the cobalt removal process, respectively.

References

[1] J.A. Sędzimir, Precipitation of metals by metals(cementation)-kinetics, equilibria, Hydrometallurgy 64 (2002) 161–167.
[2] J.H. Lee, J.M. Lee, Progress and challenges in control of chemical processes, Annual Review of Chemical and Biomolecular Engineering 5 (Jun. 2014) 383–404.
[3] G. Wang, C. Yang, H. Zhu, Y. Li, X. Peng, W. Gui, State-transition-algorithm-based resolution for overlapping linear sweep voltammetric peaks with high signal ratio, Chemometrics and Intelligent Laboratory Systems 151 (2016) 61–70.
[4] B. Zhang, C. Yang, H. Zhu, Y. Li, W. Gui, Evaluation strategy for the control of the copper removal process based on oxidation-reduction potential, Chemical Engineering Journal 284 (2016) 294–304.
[5] B. Sun, W. Gui, Y. Wang, C. Yang, M. He, A gradient optimization scheme for solution purification process, Control Engineering Practice 44 (2015) 89–103.
[6] L. Ljung, Perspectives on system identification, Annual Reviews in Control 34 (1) (2010) 1–12.
[7] B. Sun, C. Yang, Y. Wang, W. Gui, I. Craig, L. Olivier, A comprehensive hybrid first principles/machine learning modeling framework for complex industrial processes, Journal of Process Control 86 (2020) 30–43.
[8] J. Bierbrauer, Introduction to Coding Theory, Chapman and Hall/CRC, New York, 2016.
[9] D.E. Seborg, D.A. Mellichamp, T.F. Edgar, F.J. Doyle III, Process Dynamics and Control, fourth ed., John Wiley & Sons, Hoboken, NJ, 2016.
[10] L. Ljung, T. Glad, Modeling of Dynamic Systems, PTR Prentice Hall, 1994.
[11] R.E. Kalman, A new approach to linear filtering and prediction problems, Journal of Basic Engineering (1960) 35–45.
[12] L. Ljung, System Identification: Theory for the User, second ed., Prentice Hall, Upper Saddle River, NJ, 1999.
[13] Z. Ge, Review on data-driven modeling and monitoring for plant-wide industrial processes, Chemometrics and Intelligent Laboratory Systems 171 (2017) 16–25.
[14] Z. Ge, Z. Song, S.X. Ding, B. Huang, Data mining and analytics in the process industry: the role of machine learning, IEEE Access 5 (2017) 20590–20616.
[15] G. Buzzi-Ferraris, F. Manenti, Kinetic models analysis, Chemical Engineering Science 64 (5) (2009) 1061–1074.
[16] C.C. Pantelides, J. Renfro, The online use of first-principles models in process operations: review, current status and future needs, Computers & Chemical Engineering 51 (2013) 136–148.

Part II

Modeling and optimal control of the copper removal process

Chapter 3

Kinetic modeling of the competitive-consecutive reaction system

Contents

3.1 Process description and analysis

Copper ions are the main impurity in the leaching solution. The concentration of copper ions in the leaching solution is much higher than that of the other impurities to be removed, e.g., cobalt and nickel [1][2][3]. The impurity removal reaction is essentially a replacement reaction between zinc powder and impurity metal ions. Electrons are transferred from zinc powder to impurity ions. Therefore, impurity ions obtain the electrons and are reduced. The reaction product contains impurity cemented on the surface of zinc powder. The difference between the equilibrium potential of anode and cathode reactions determines the driving force of the replacement reaction. The oxidation of zinc is an anode reaction, as shown in (3.1). The reduction of impurity ions is a cathode reaction. Reactions (3.2) to (3.4) present the cathode reactions in the copper removal process and the cobalt removal process. It can be observed that copper has a larger equilibrium potential difference with zinc than the other impurity metals. In addition, there is no overvoltage phenomenon in copper deposition. Therefore, the replacement reaction between zinc and copper ions has a large reaction impetus, which indicates copper can be easily replaced by zinc compared with cobalt. In practice, copper is the first impurity ion removed in the purification process. We

Modeling, Optimization, and Control of Zinc Hydrometallurgical Purification Process
https://doi.org/10.1016/B978-0-12-819592-5.00013-2

have the following reactions:

$$Zn \rightarrow Zn^{2+} + 2e^-, \qquad E_e = -0.763\ V, \tag{3.1}$$

$$Cu^{2+} + 2e^- \rightarrow Cu, \qquad E_e = 0.337\ V, \tag{3.2}$$

$$Co^{2+} + 2e^- \rightarrow Co, \qquad E_e = -0.277\ V, \tag{3.3}$$

$$Ni^{2+} + 2e^- \rightarrow Ni, \qquad E_e = -0.250\ V. \tag{3.4}$$

As copper ions can be easily removed by zinc powder compared with cobalt, if sufficient zinc powder is added in the copper removal process, deep purification of copper ions can be achieved. However, copper ion is not only an impurity. It can also act as catalyst together with arsenic trioxide in the subsequent cobalt removal process. More specifically, they can react with arsenic trioxide and cement on the zinc powder surface, providing a reaction surface for cobalt cementation. In addition, as the equilibrium potential difference between copper and zinc is large, by leaving some copper ions, cobalt cementation can be further promoted. Therefore, the operation objective of the copper removal process is not to reduce the copper ion concentration such that it is below a certain threshold, i.e., $0 < C \leqslant C_{up}$; the objective is to control the copper ion concentration such that it is within a reasonable range $[C_{low}, C_{up}]$ bounded by an upper limit C_{up} and a lower limit C_{low} above zero. Therefore, as the first stage in purification process, the performance of copper removal not only affects the technical and economical indices of the copper removal process, but also affects the inlet conditions and stability of the following cobalt removal process.

Conventionally, the copper removal process is composed of several continuous stirred tank reactors and other auxiliary equipments, e.g., a surge tank and a pressure filter, etc., as shown in Fig. 3.1 [4][5][6][7]. In each reactor, the copper ions are removed according to the following reaction:

$$CuSO_4 + Zn \rightarrow ZnSO_4 + Cu \downarrow . \tag{3.5}$$

The copper ions or copper sulfate in the leaching solution is reduced by zinc powder and precipitated. In some plants, cadmium ions are removed together

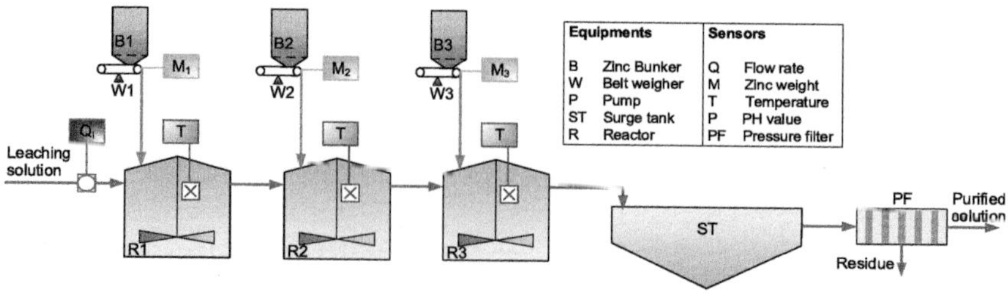

FIGURE 3.1 Conventional copper removal process.

with copper ions in the cobalt removal process. According to reaction (3.5), under ideal stoichiometric conversion conditions, 1 mol of zinc powder can remove 1 mol of copper ions. However, the theoretical stoichiometric conversion is difficult to obtain in practical production environments with dynamic and uncertainty fluctuations. Instead, more than 1 mol zinc is usually required to remove 1 mol copper ions. The unreacted zinc therefore remains in the residue.

In the ADL process, the flow path of the copper removal process is improved. As shown in Fig. 3.2, the feeding is the leaching solution. The main part of the improved copper removal process includes several cascaded continuous stirred tank reactors and a thickener. The copper removal reaction takes place in the reactors. After retention in the cascaded reactors, the purified solution flows to a thickener in which solution and the solid resultants are separated. The supernate flows to the next process. Some of the solid resultant is recycled to reactor #1 as underflow. The remaining solid resultant is sent for residue processing.

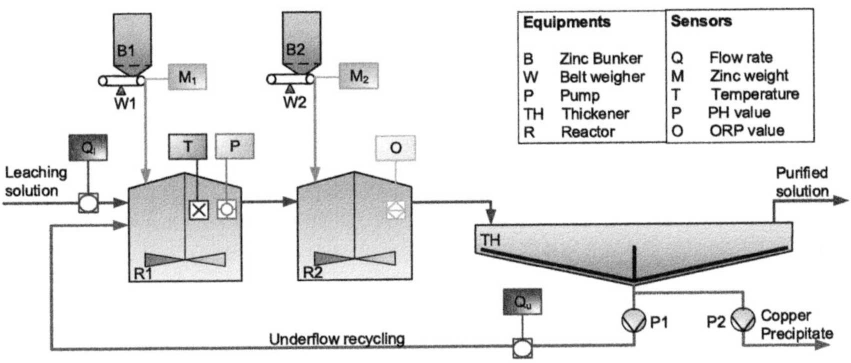

FIGURE 3.2 Copper removal process with recycling.

From Figs. 3.1 and 3.2, the difference between the conventional copper removal process and the improved copper removal process can be observed. In the improved process, part of the thickener underflow containing copper residue is returned to the first reactor. The recycled copper residue can react with copper sulfate according to the following reaction, in which the copper ions are precipitated as cuprous oxide [8]:

$$CuSO_4 + Cu + H_2O \rightarrow Cu_2O \downarrow + H_2SO_4. \tag{3.6}$$

This indicates that one more reaction path for copper removal is generated by recycling the underflow. Therefore, by using the improved copper removal technology, the copper ions are deposited in two different ways [9]:

- Reaction (3.5): ionic copper is replaced by zinc powder and precipitated as metallic copper.
- Reaction (3.6): ionic copper reacts with the recycled metallic copper and precipitates as cuprous oxide.

By combining the two reactions, the overall copper removal reaction can be described as:

$$2CuSO_4 + Zn + H_2O \rightarrow ZnSO_4 + Cu_2O \downarrow + H_2SO_4. \qquad (3.7)$$

Considering ideal stoichiometric conversion conditions, it can be indicated that in reaction (3.5), 1 mol zinc is required to remove 1 mol copper ions. However, in reaction (3.7), only 1/2 mol zinc is required to remove 1 mol copper ions. Besides higher zinc powder utilization efficiency, the improved copper removal technology possesses the following advantages:

(i) The recycled thickener underflow contains not only copper residue, but also unreacted zinc powder. Therefore, the unreacted zinc particles are also recycled and reused, which can further reduce zinc powder consumption.
(ii) The precipitated cuprous oxide enables the removal of chloride existing in the solution.

The operation of the copper removal process is to adjust the zinc powder dosage of each reactor and the reaction conditions (temperature, pH), such that the outlet copper ion concentration of the last reactor is within the desired range $[C_{low}, C_{up}]$. However, the determination of zinc powder dosage of each reactor encounters the following two problems.

On the one hand, the change in the flow path not only brings advantages like higher copper removal efficiency and zinc powder utilization efficiency, but also increases the complexity of the process, which can be explained from two aspects:

(i) More complex stoichiometric conversion: In the improved copper removal technology, there exist two main reactions, i.e., copper cementation (3.5) and cuprous oxide precipitation (3.6). On the one hand, ionic copper is consumed in both reactions. On the other hand, as a product of reaction (3.5), the metallic copper is also the reactant of reaction (3.6). This indicates that these two reactions compete for ionic copper. However, the proceeding of reaction (3.5) can promote reaction (3.6). Therefore, these correlated two reactions form a "competitive-consecutive" reaction system. As the stoichiometric conversion rates between zinc powder and copper ions in these two reactions are different, the overall stoichiometric conversion is complicated.
(ii) More complex process dynamics: Due to the correlation between the two main reactions, the outlet copper ion concentration and the mass ratio between the metallic copper and cuprous oxide keep stable if a dynamic balance between copper cementation (6.4) and cuprous oxide precipitation (6.3) is achieved. A small change of the reaction or inlet conditions would brake the equilibrium of the "competitive-consecutive" reaction system, resulting in fluctuations in the outlet copper ion concentration.

On the other hand, the variation of the inlet and reaction conditions of the copper removal process is complex. Under complex working conditions, unreasonable operations could result in failure to meet the technical requirement on outlet copper ion concentration, so the copper ion concentration in the feeding solution of the subsequent cobalt removal process is overly high, which then influences the cobalt removal performance. The complexity of working conditions includes:

(i) Fluctuating feeding conditions: The supply-side volatility gives rise to the variation of copper content in zinc concentrate, resulting in fluctuating feeding conditions, i.e., the inlet copper ion concentration is not stable.
(ii) Delayed chemical assay [10][11][12]: In the copper removal process, due to the high temperature and erosion of zinc sulfate solution, the online element analyzer is hard to maintain. Therefore, the copper ion concentration is usually obtained using time-consuming chemical assays. The fluctuation in the inlet copper ion concentration is difficult to be detected in real-time.
(iii) Limited measurements: The measurements of copper ion concentrations and the composition analysis of the underflow are insufficient. In the inlet solution, the concentrations of most metal ions are detected. However, as the concentrations of other elements are relatively low, in the purification solution (outlet solution of the last reactor), only the copper ion concentration is measured. However, other impurities also consume zinc powder and affect copper removal, e.g., an excessively high ferrous ion concentration can reduce the copper removal efficiency.

Theoretically, the improved copper removal process can reduce zinc powder consumption. However, due to the complexity of the improved copper removal process, human operators usually use excessive amounts of zinc powder to guarantee that the outlet copper ion concentration is lower than the technical threshold, which is a waste of valuable materials. Therefore, the advantage of the improved technology is not fully utilized. For economic and stable production, it is necessary to study the reaction kinetics of the copper removal process, and to obtain a comprehensive model covering the process dynamics under different working conditions.

3.2 Kinetics of copper removal reactions

In the remainder of this chapter, a kinetic model of the copper removal process is constructed to describe the process dynamics under different working conditions. The copper removal process includes two main reactions, i.e., copper cementation and cuprous oxide precipitation. The kinetic models of these two reactions are first derived separately. Then, the kinetics of the overall "competitive-consecutive" reaction system is studied, and the model structure of the reaction system is determined [13]. The model parameters under different working conditions are identified based on an intelligent optimization approach. To start with, in this section, the main influencing factors and reaction mechanism of the two main reactions are analyzed.

3.2.1 Influencing factor analysis

The copper removal process is a solid-liquid phase, noncatalyst, and multiple-reaction process. The copper removal performance is influenced by numerous factors, including temperature, reaction time, pH, composition of the leaching solution [12], zinc powder dosage, solid content of underflow, etc.

3.2.1.1 *Temperature*

According to the Arrhenius equation, the reaction rate of copper removal can be increased by raising the solution temperature. On the one hand, the increase in temperature can decrease the concentration polarization and electrochemical polarization of electrode processes, which provides more reaction impetus for the two main reactions. On the other hand, the two main reactions are exothermic. Although the increase in temperature can accelerate the diffusion of reactant particles, an overly high temperature is not beneficial to copper removal from the perspective of thermodynamics. Therefore, in practice, the temperature of the copper removal process is stabilized around a certain value, depending on the production environment of the plant.

3.2.1.2 *Reaction time*

As copper ions can be easily replaced by zinc powder, the reaction rate is high. If the retention time of the solution in the reactor is too long, copper ions can be removed to a very low concentration, which cannot facilitate the following cobalt removal process. However, if the retention time is too short, the precipitation of copper ions is insufficient. In practice, the retention time is controlled by limiting the flow rate of inlet solution in a reasonable range.

3.2.1.3 *pH*

In the leaching process, sulfate acid is added to react with zinc calcine. The pH values of each leaching reactor are important technical indices of the leaching process. The leaching solution is faintly acid. So in the copper removal process, a small portion of the added zinc powder would react with the hydrogen ions in the solution. Therefore, when the pH is overly low, the reduction of hydrogen ions could intensify, which competes with valuable zinc powder in the copper removal reaction. When the pH is overly high, basic zinc sulfate could be generated. It attaches on the surface of zinc powder particles and hinders the electron transfer between zinc powder and copper ions. Copper ions therefore cannot be reduced, i.e., Cu cannot be produced. So the cuprous oxide precipitation reaction lacks reactant. This indicates that if the pH is overly high, the reaction rate of the two main reactions could decrease.

3.2.1.4 *Composition of leaching solution*

Besides the two main reactions, there exist side reactions introduced by other impurities in the leaching solution. Some of these impurities could compete with copper ions for zinc powder, e.g., ferrous ions. Therefore, if the concentration of these impurities is high, the zinc powder consumption will increase. On the other hand, if the concentration of copper ions in the leaching solution is overly low, the outlet copper ion concentration of the last reactor may be lower than the lower limit C_{low}.

3.2.1.5 Zinc powder dosage

Zinc powder dosage is the most important manipulated variable of the copper removal process. Overly low or overly high zinc powder dosages could result in unsatisfactory copper removal performance. The physical properties of zinc powder also affect copper removal. The copper removal reaction takes place on the surface of zinc powder particles. Therefore, the shape and size of zinc powder particles affect the reaction area of copper removal. Zinc powder particles with small size have a larger surface area. However, too fine zinc powder particles could hydrolyze in the solution before they can react with copper ions.

3.2.1.6 Solid content of underflow

The solids in the underflow contain the reaction product of reaction (3.5). It serves as the seed crystal of reaction (3.6). If the content of solids in the underflow is overly low, the amount of seed crystal is insufficient. So copper ions are mainly removed by (3.5) compared with reaction (3.6). Considering the stoichiometric conversion between zinc powder and copper ions in these two reactions, more zinc powder is required. If the content of solids in the underflow is overly high, the solution becomes sticky, which also affects the copper removal efficiency.

To summarize, all the above factors should be controlled within appropriate ranges. Among these factors, solution temperature, pH, zinc powder particle size, and solid content of underflow are determined in the trial operation phase and kept in reasonable ranges in most cases. In the production phase, the flow rate and copper ion concentration of the inlet solution fluctuate and become the main influencing factors.

3.2.2 Copper cementation kinetics

From the microcosmic perspective, reaction (3.5) mainly consists of three stages:

(i) External diffusion: The liquid reactant ($CuSO_4$) arrives on the external surface of the solid reactant (zinc powder particles) via a retention membrane.
(ii) Internal diffusion: The liquid reactant further diffuses from the external surface to the internal surface via a porous channel.
(iii) Surface chemical reaction: The liquid reactant reacts with the solid reactant.

Among the above three stages, the rate of stage (3) is the lowest. So the controlling step of reaction (3.5) is the surface chemical reaction.

In reaction (3.5), the copper ions obtain electrons from zinc powder and are cemented on the zinc particles. The cementation reaction can be considered as a surface-controlled first-order reaction whose reaction rate can be formulated as

$$r_1 = -KC = -\frac{K_D S_A}{V} C_{Cu^{2+}}, \tag{3.8}$$

where r_1 means the reaction speed, K means the reaction rate constant which varies with different reaction conditions, C means the reactant concentration, $C_{Cu^{2+}}$ is the copper concentration, K_D is the mass transfer coefficient, V is the reactor's volume, and S_A is the surface area for reaction (3.5), which is determined by the quantities of newly added and recycled zinc powder [14]. In addition, zinc powder particles with a smaller size have a larger specific surface region compared with particles with a larger size. Therefore, the zinc powder particles' size distribution also affects the surface area for copper cementation.

To estimate the total reaction surface area, it is necessary to obtain the size of zinc powder particles. In practice, the particle size is obtained using a sieve with adjustable screen size. Therefore, by using a sieve, the tested zinc powder particles with different size can be classified into different categories. Assuming that there are N_{size} types of particle size, if the weight of particles sieved using screen size S_i is m_i, then its weight fraction is

$$\alpha_i \approx \frac{m_i}{\sum_{i=1}^{N_{size}} m_i} = \frac{m_i}{M}, \tag{3.9}$$

where $M = \sum_{i=1}^{N_{size}} m_i$.

Considering the particles sieved using screen size S_i and assuming the shape of the zinc powder particles is spherical, their total surface area is

$$S_i^A = N_{total} \alpha_i \pi d_i^2, \tag{3.10}$$

where N_{total} is the total number of tested zinc particles and d_i is the average particle diameter, i.e., the average value of the limits of the screen size interval. Its total volume is

$$V^P{}_i = \frac{1}{6} N_{total} \alpha_i \pi d_i^3. \tag{3.11}$$

Because the density ρ is the same for the zinc powder particles with different sizes, the surface of all the tested zinc powder particles is

$$W = \rho V^P = \rho \sum_{i=1}^{N_{screen}} \frac{1}{6} N_{total} \alpha_i \pi d_i^3. \tag{3.12}$$

Therefore, the total reaction surface area can be estimated using the weight and density of zinc powder particles and the weight fractions of the particles with different sizes:

$$S^{A} = \frac{6\sum_{i=1}^{n} \alpha_i d_i^2}{\rho \sum_{i=1}^{n} \alpha_i d_i^3} W. \tag{3.13}$$

In the improved copper removal process, the zinc powder particles in a reactor include newly added zinc powder as well as zinc powder particles in the recycled underflow. Therefore, the surface area of zinc powder for copper cementation is

$$S = \frac{6\sum_{i=1}^{n} \alpha_i d_i^2}{\rho \sum_{i=1}^{n} \alpha_i d_i^3} (g_{Zn} + \gamma_{Zn} g_{under}), \tag{3.14}$$

where g_{Zn} is the quantity of fresh zinc powder in the zinc bin, g_{under} is the quantity of recycled solids, and γ_{Zn} is the zinc powder's weight fraction in the recycled solids. To sum up, the reaction rate of copper cementation can be formulated as

$$r_1 = -\frac{6K_D (g_{Zn} + \gamma_{Zn} g_{under}) \sum_{i=1}^{n} \alpha_i d_i^2}{V\rho \sum_{i=1}^{n} \alpha_i d_i^3} C_{Cu^{2+}}. \tag{3.15}$$

3.2.3 Cuprous oxide precipitation kinetics

According to reaction (3.6), after copper cementation, copper ions react with cemented copper and form cuprous oxide. This reaction can increase the copper removal efficiency. As reaction (3.5), reaction (3.6) can be categorized as a noncatalytic fluid-solid reaction. The microscopic reaction stages of reaction (3.6) are also the same as those of reaction (3.5). However, as the reaction environment is time-varying and the order of the reaction rates of these three stages is not constant, it is difficult to determine the controlling step of reaction (3.6). Therefore, the influence of all three stages on the rate of reaction (3.6) is considered.

The shrinking core model (SCM) is a widely applied noncatalytic fluid-solid reaction model (Fig. 3.3) [15,16]. In this model, the solid reactant is assumed to be ideal spherical particles. The chemical reaction takes place on the interface of the reaction product layer and the unreacted solid core. The reaction spreads from the reaction product to the unreacted solid. Therefore, the unreacted solid shrinks towards the core. SCM is a suitable model to describe the noncatalytic reaction process between solid and liquid reactants.

According to the SCM theory, reaction (3.6) contains three steps:

(i) External diffusion: Cu^{2+} diffuses via the liquid film surrounding the elemental copper.

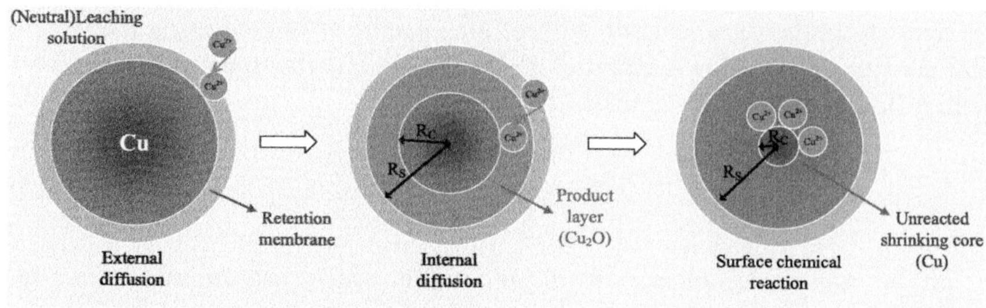

FIGURE 3.3 Shrinking core model of reaction (3.6).

(ii) Internal diffusion: Cu^{2+} diffuses to the internal surface of elemental copper particles (through the product layer).

(iii) Surface chemical reaction: reaction (3.6) takes place, increasing the thickness of the Cu_2O reaction product layer. The elemental copper particle, which acts as an unreacted core, shrinks as the reaction proceeds.

In order to determine the controlling step of the reaction, the reaction rates of all the three stages are studied and compared. The reaction rate of the external diffusion stage is

$$r_{2,1} = -4\pi R_S^2 V^{-1} k_G (C_{Cu^{2+}g} - C_{Cu^{2+}s}), \tag{3.16}$$

where $r_{2,1}$ means the reaction rate of the external diffusion stage, R_S means the particle radius, k_G means the mass transfer coefficient, $C_{Cu^{2+}g}$ means the copper concentration in the bulk phase, $C_{Cu^{2+}s}$ means the copper concentration of the copper particle at the outer surface, and V is the volume of the particles.

In the internal diffusion stage, the liquid reactants ($CuSO_4$ and H_2O) diffuse to the internal surface of elemental copper particles through the Cu_2O layer. Its reaction rate is

$$r_{2,2} = -4\pi R_S^2 V^{-1} D_e \left(\frac{dC_{Cu^{2+}}}{dR}\right)_{R=R_C}, \tag{3.17}$$

where $r_{2,2}$ means the reaction rate of the internal diffusion stage, D_e means the effective diffusivity of $CuSO_4$ through the Cu_2O layer, and R_C means the distance from the core to the reaction surface.

In the surface chemical reaction stage, the precipitation happens on the surface of the unreacted copper particles, which is a first-order reaction. Its reaction rate is

$$r_{2,3} = -4\pi R_C^2 V^{-1} k_{surf} C_{Cu^{2+}c}, \tag{3.18}$$

where $r_{2,3}$ means the reaction rate of the surface chemical reaction stage, k_{surf} is the reaction rate constant, and $C_{Cu^{2+}c}$ means the copper concentration on the reaction surface.

Generally, the controlling step of a reaction is the microscopic stage with the lowest rate, which determines the rate of the reaction. However, as explained above, it is difficult to quantify the reaction rates of the three stages. Therefore, none of the three stages are ignored, and the overall reaction rate is formulated as the sum of the rates of three stages:

$$r_2 = -4\pi R_S^2 V^{-1} C_{Cu^{2+}g}\left[\frac{1}{k_G} + \frac{R_S(R_S - R_C)}{R_C D_e} + \frac{R_S^2}{R_C^2 k_{surf}}\right]. \tag{3.19}$$

Assuming that the solid particles of reaction product are ideal spheres and only contain Cu_2O and Cu and that the product layer and the copper core are homogeneous, the weight fraction ratio of Cu_2O in the precipitate, which is measured routinely, can be approximated as

$$\theta = \frac{W_{Cu_2O}}{W_{Cu_2O} + W_{Cu}} = \frac{(R_S^3 - R_C^3)\rho_{Cu_2O}}{R_C^3\rho_{Cu} + (R_S^3 - R_C^3)\rho_{Cu_2O}} \approx 1 - \frac{R_C^3}{2R_S^3 - R_C^3}, \tag{3.20}$$

where W_{Cu_2O} and W_{Cu} are the weights of cuprous oxide and metallic copper in the solid and ρ_{Cu_2O} and ρ_{Cu} are the densities of cuprous oxide and metallic copper, respectively. Then the unreacted core's radius and the overall reaction rate can be expressed as

$$R_C - \sqrt[3]{\frac{2-2\theta}{2-\theta}} R_S = \eta R_S, \tag{3.21}$$

where $\eta = [(2-2\theta)/(2-\theta)]^{-1/3}$.

Substituting (3.21) into (3.19), the overall reaction rate of (3.6) can be expressed as

$$r_2 = -4\pi R_S^2 V^{-1} C_{Cu^{2+}g}\left[\frac{1}{K_G} + \frac{(\eta - 1)R_S}{D_e} + \frac{\eta^2}{k_{surf}}\right], \tag{3.22}$$

where R_S, K_G, D_e, and k_{surf} are to be identified.

3.3 Modeling of the competing reactions system

3.3.1 Model structure determination

In the improved copper removal process, the copper cementation and cuprous oxide precipitation reactions take place parallelly. Copper ions are consumed in both reactions and one of the products of the former reaction becomes a reactant of the latter reaction. These two reactions form a "competitive-consecutive" reaction system (CCRS):

$$A + B \xrightarrow{k_1} C\downarrow + D, \tag{3.23}$$

$$A + C + E \xrightarrow{k_2} R\downarrow + H, \tag{3.24}$$

where k_1 with k_2 are the rate constants of the two reactions. In the context of copper removal, the symbols A, B, C, D, E, R, and H stand for $CuSO_4$, Zn, Cu, $ZnSO_4$, H_2O, Cu_2O, and H_2SO_4, respectively.

These two reactions follow the first-order reaction kinetics. Their reaction rates can be expressed as

$$r_i = -k_i c_{Cu^{2+}}, \tag{3.25}$$

where r_i and k_i are the reaction rate and rate constant and $i = 1, 2$. According to the results in Sections 3.2.2 and 3.2.3,

$$k_1 = \frac{6K_{\mathrm{D}}(g_{\mathrm{Zn}} + \gamma_{\mathrm{Zn}} g_{\mathrm{under}}) \sum_{i=1}^{n} \alpha_i d_i^2}{V_\rho \sum_{i=1}^{n} \alpha_i d_i^3}, \tag{3.26}$$

$$k_2 = 4\pi R_{\mathrm{S}}^2 V^{-1} \left[\frac{1}{K_{\mathrm{G}}} + \frac{(\eta - 1) R_{\mathrm{S}}}{D_e} + \frac{\eta^2}{k_{surf}} \right]. \tag{3.27}$$

However, the reaction mechanism of a single reaction cannot describe the overall reaction system. It is required to analyze the influence of these two reactions on the copper removal performance. It can be observed from the two reactions that the two reactions compete with each other as they both use Cu^{2+} as reactant. In addition, reaction (3.5) generates reactant for reaction (3.6). Therefore, these two reactions hold a "consecutive reaction" relationship. According to the independence principle of chemical reactions, if the orders of two reactions are equal, then the consumption rate of common reactant (Cu^{2+}) equals the sum of each reaction. However, in consecutive reactions, the consumption rate of the related material (Cu) relies not only on the reaction order, but also on the reaction constant. When the reaction orders are the same, however their reaction constant are different in order of magnitude, then the consumption rate of the related material relies on the reaction with the smaller reaction constant. When the reaction orders are the same and the reaction constants are similar, then the consumption rate of the related material equals the reaction rate of consumption minus the reaction rate of production. To sum up, the reaction rates of each reactant/product of the CCRS are:

$$\begin{aligned}
r_{Cu^{2+}} &= -r_1 - r_2, \\
r_{Zn} &= -r_1, \\
r_{Zn^{2+}} &= r_1, \\
r_{H_2O} &= -r_2, \\
r_{Cu_2O} &= r_2, \\
r_{H^+} &= r_2,
\end{aligned}$$

$$r_{Cu} = \begin{cases} r_2, & k_1 >> k_2, \\ -r_1, & k_1 << k_2, \\ -r_1 + r_2, & \text{else}, \end{cases}$$

where $r_{Cu^{2+}}$, r_{Zn}, $r_{Zn^{2+}}$, r_{H_2O}, r_{Zn}, r_{Cu_2O}, r_{H^+}, and r_{Cu} are the reaction rates of Cu^{2+}, Zn, Zn^{2+}, H_2O, Zn, Cu_2O, H^+, and Cu, respectively.

Outlet copper ion concentrations of each reactor are the KPIs of the copper removal process. Considering a single reactor, the reaction rate of Cu^{2+} can be derived according to the above result:

$$r_{\mathrm{Cu}^{2+}} = -\left\{ \frac{6K_{\mathrm{D}}(g_{\mathrm{Zn}} + \gamma_{\mathrm{Zn}} g_{\mathrm{under}}) \sum_{i=1}^{n} \alpha_i d_i^2}{\rho \sum_{i=1}^{n} \alpha_i d_i^3} + 4\pi R_{\mathrm{S}}^2 \left[\frac{1}{K_{\mathrm{G}}} + \frac{(\eta - 1) R_{\mathrm{S}}}{D_e} + \frac{\eta^2}{k_{surf}} \right] \right\}, \tag{3.28}$$

where $r_{\mathrm{Cu}^{2+}}$ is the consumption rate of copper ion concentration.

The dynamics of the copper removal reactor can be described using the classical CSTR model [17]. Then according to the above kinetic study and mass balance, the dynamics of the copper removal reactors follow

$$V\dot{C}_{\mathrm{Cu}^{2+},1} = QC^{0}_{\mathrm{Cu}^{2+},1} - (1 + Q_{\mathrm{uf}})QC_{\mathrm{Cu}^{2+},1} - Vr_{\mathrm{Cu}^{2+}}, \tag{3.29}$$

$$V\dot{C}_{\mathrm{Cu}^{2+},2} = (1 + Q_{\mathrm{uf}})QC^{0}_{\mathrm{Cu}^{2+},2} - (1 + Q_{\mathrm{uf}})QC_{\mathrm{Cu}^{2+},2} - Vr_{\mathrm{Cu}^{2+}}, \tag{3.30}$$

where $\dot{C}_{\mathrm{Cu}^{2+},\mathrm{i}}$, $i = 1, 2$, is the reaction rate of copper ions in the ith reactor, $C^{0}_{\mathrm{Cu}^{2+},\mathrm{i}}$, $i = 1, 2$, is the inlet copper ion concentration of the ith reactor, and Q and Q_{uf} are the flow rates of zinc sulfate solution and the returned underflow, respectively. The details of model parameters are summarized in Table 3.1.

3.3.2 Model parameter identification

The dynamics of the copper removal process have various influencing factors. With different combinations of inlet conditions and reaction conditions, the process dynamics exhibit different features. The different process dynamics are presented by the difference in model parameters. Therefore, a mathematical model with only one set of model parameters is insufficient to comprehensively describe the process dynamics under different working conditions. In practice, a mathematical model with fixed structure and model parameters changing with the working conditions is usually adopted. Therefore, to guarantee the accuracy of the kinetic model, the model parameters under each working condition should be identified. However, it is difficult to recognize different working conditions according to artificial experience. On the other hand, the working load of classifying the large volume of data samples artificially is heavy, and prone to classification errors. Therefore, it is necessary to design an adaptive parameter identification approach which automatically classifies the data samples.

This section presents a model parameter identification framework for processes with various working conditions based on data features. As illustrated

TABLE 3.1 Parameters in the kinetic model.

Parameter	Physical meaning
V	Volume of reactor
Q	Flow rate of leaching solution
Q_{uf}	Flow rate of recycled underflow
$\dot{C}_{Cu^{2+},i}$	Reaction rate of copper ions in the ith reactor
$C^0_{Cu^{2+},i}$	The inlet copper ion concentration of the ith reactor
$C_{Cu^{2+},i}$	The copper ion concentration in the ith reactor
$r_{Cu^{2+}}$	Consumption rate of copper ions
r_1	Reaction rate of copper cementation
r_2	Reaction rate of cuprous oxide precipitation
$r_{2,1}$	Reaction rate of external diffusion in cuprous oxide precipitation
$r_{2,2}$	Reaction rate of internal diffusion in cuprous oxide precipitation
$r_{2,3}$	Reaction rate of surface chemical reaction in cuprous oxide precipitation
$C_{Cu^{2+},g}$	The copper ion concentration in the bulk phase
$C_{Cu^{2+},s}$	The copper ion concentration at the outer surface of the copper particles
$C_{Cu^{2+},c}$	The copper ion concentration on the reaction surface
R_S	Particle radius
R_C	Distance from the core to the reaction surface
k_G	Mass transfer coefficient of external diffusion of reaction (3.6)
D_e	Effective diffusivity of internal diffusion of reaction (3.6)
k_{surf}	Surface chemical reaction rate constant of reaction (3.6)
K_D	Mass transfer coefficient of reaction (3.5)
S_A	Surface area of reaction (3.5)
ρ	Density of zinc powder particles (3.5)
g_{Zn}	Weight of added zinc powder
g_{under}	Weight of recycled solids
γ_{Zn}	Weight fraction of zinc powder in the recycled solids

in Fig. 3.4, the model parameter identification framework is composed of three parts, i.e., sample classification, sample balancing, and parameter estimation. These three parts correspond to the three stages of the parameter identification process.

Stage 1 The first stage is the sample classification stage. In this stage, the data samples collected via a data collection campaign are divided into several groups according to their data features. As the model parameters take different values under different working conditions, the process variables associated with the process model present different data features. Therefore, the classified data samples correspond to different working conditions, which can be used to identify model parameters of different working conditions.

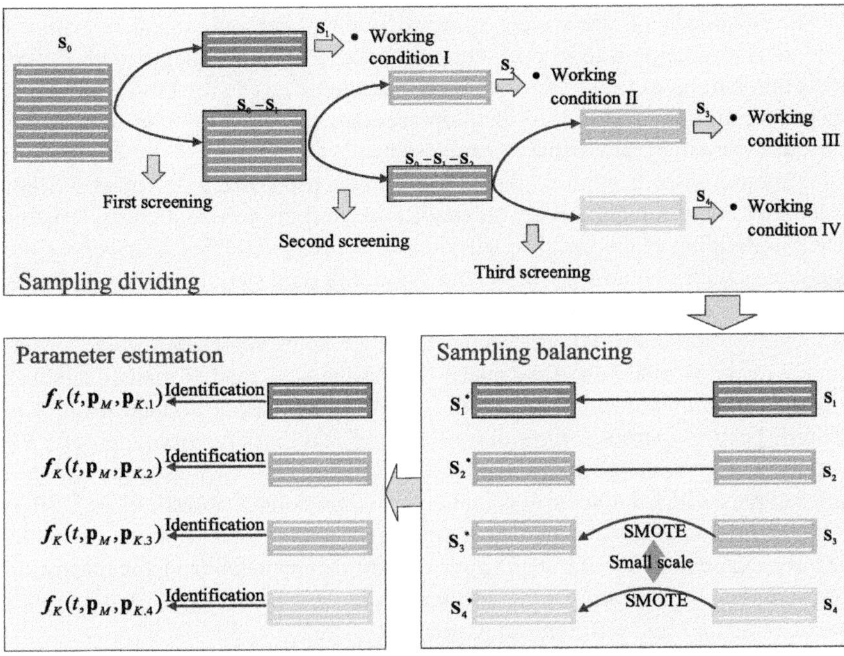

FIGURE 3.4 Model parameter identification framework.

Stage 2 The second stage is the sample balancing stage. As the numbers of data samples classified into different categories are not balanced, the data samples belonging to some working conditions are insufficient to identify the model parameters accurately. Therefore, in this stage, new data samples are generated and added to the working conditions with fewer data samples to eliminate the size imbalance problem.

Stage 3 The third stage is the parameter estimation stage. In this stage, intelligent optimization algorithms are applied to identify the unknown model parameters of each working condition from the classified data samples.

3.3.2.1 *Data sample labeling and classification*

By numerical integration of the dynamic model of the copper removal process, the formulation of the real-time value of the outlet copper concentration can be described as

$$\mathbf{C}_{\mathbf{Cu}^{2+}} = \mathbf{MODEL}(t, \mathbf{p}_{\mathrm{cn}}, \mathbf{p}_{\mathrm{mp}}), \tag{3.31}$$

where $\mathbf{C}_{\mathbf{Cu}^{2+}}$ is the vector of outlet copper ion concentrations, **MODEL** is a vector of functions which is the integral form of the dynamic model, t is time, $\mathbf{p}_{\mathrm{cn}}$ is a vector of independent variables ($C^0_{\mathrm{Cu}^{2+},1}(t)$, $C^0_{\mathrm{Cu}^{2+},2}(t)$, Q, q, γ_{Zn}, g_{Zn}, g_{under}, α_i, d_i, ρ, V, R_{S}, and η), and $\mathbf{p}_{\mathrm{mp}}$ is a vector of model parameters to be identified (K_{D}, k_{G}, D_e, and k).

The dynamics of the copper removal process are influenced by various factors. The reaction rate in each copper removal reactor changes with different combinations of these factors which indicate different working conditions. Therefore, to increase the identification accuracy, the data samples are first labeled and classified into different categories.

Without loss of generality, the dynamics of a process are a synthetic result of its inlet conditions, the reaction conditions, and the inherent characteristics of the underlying reactions. The variation of process variables is driven by the essential process dynamics. On the one hand, the data samples associated with different working conditions present different features. On the other hand, the data samples associated with the same working condition present similar consistency with a potential model parameter set. As the workload of manual labeling is heavy, an automatic sample labeling approach based on the data features is desired. For this purpose, a condition labeling and classification approach according to the model performance is proposed. In each identification run, the data samples with relative errors smaller than a predefined threshold are treated as belonging to the same working condition. Then, in the next run, the data samples associated with another working condition are extracted from the remaining data samples. By repeating the identification procedure, all of the data samples can be classified into different categories.

The detailed steps of the sample labeling and classification approach include:

Step 1 Randomly extract a sample set $\mathbf{S}_0 = \{s_i | i = 1, ..., m_0\}$ from the collected data samples. Determine the value of the precision threshold $\upsilon_{\text{precision}}$ and the size threshold υ_{size}.

Step 2 Identify the model parameters $\mathbf{P}_0 = \{K_{D0}, k_{G0}, D_{e0}, k_0\}$ by minimizing the total error between the model output and the real output of all the data samples in $\mathbf{S}_0$. The model with parameter set $\mathbf{P}_0$ can then be expressed as $M(\mathbf{P}_0)$.

Step 3 Calculate the relative errors for all the data from $\mathbf{S}_0$:

$$e_i^0 = |(y_i^0 - c_i)/c_i|, \quad i = 1, 2, ..., m_0, \tag{3.32}$$

where y_i^0 is the model output of the ith data sample (the model output is calculated by using $\mathbf{P}_0$ as model parameters) and c_i is the actual output of the ith data sample.

Step 4 Screen data samples whose error is less than $\upsilon_{\text{precision}}$ out of $\mathbf{S}_0$ to form a new data set $\mathbf{S}_1$, which is labeled as working condition I: $\mathbf{S}_1 = \left\{\mathbf{s}_i | i = 1, ..., m_1, e_i^0 < \upsilon_{\text{precision}}, 0 < m_1 < m_0\right\}$.

Step 5 If the size of remaining data samples is larger than $\upsilon_{\text{size}} \times m_0$, then repeat *Step* 2 to *Step* 4 for further screening.

Step 6 If the number of remaining data is less than or equal to $\upsilon_{\text{size}} \times m_0$, then the remaining data set $\mathbf{S}_{\text{CN}}$ is considered as belonging to the last working condition. Therefore, the data sample set can be automatically divided into different working conditions $\{\mathbf{S}_1, \mathbf{S}_2, ..., \mathbf{S}_{\text{CN}}\}$.

3.3.2.2 Data sample balancing

In the sample labeling and classification stage, the data samples are classified into different types based on the performance in parameter identification. However, the data samples associated with different working conditions are not balanced in size. This is due to the fact that there exist some uncommon working conditions, e.g., extremely high or low amounts of zinc powder, and too high or too low inlet copper ion concentrations. The data samples associated with these working conditions are rare and insufficient to accurately estimate the unknown model parameters, whereas the data samples associated with common working conditions are sufficient. The size imbalance affects the identification accuracy.

To address these problems, the numbers of data samples associated with different working conditions should be balanced before model parameter identification. To this end, the synthetic minority oversampling technique (SMOTE) is applied to generate sufficient data samples for the working conditions with rare data samples [18]. The SMOTE algorithm generates new samples with reference to existing minority samples by keeping high similarity in data features. For the object data sample set $\mathbf{S}_{obj}$, we first find the K-nearest neighbors $X_{I,K} = \{x_{i,1}, x_{i,2}, ..., x_{i,K}\}$ for each sample x_i, where $X_{I,K}$ are a set of K data samples, whose Euclidean distances from x_i are of the smallest magnitude. Then, the new data samples can be generated as

$$x_{new.j} = x_i + (x_{i,j} - x_i) \times b, \quad j = 1, 2, ..., K, \tag{3.33}$$

where $b \in [0, 1]$ is a random number.

To sum up, the data sample balancing approach has four steps:

Step 1 Obtain the sample number of all the working conditions ($m_1, ..., m_2, ..., m_{N_w}$) and the number of working conditions N_w, and set the ratio threshold v_{ratio}.

Step 2 Calculate the sample ratio, $r_i = m_i / \sum_{j=1}^{N_w} m_j \quad (i = 1, 2, ..., N_w)$, for each working condition.

Step 3 If the sample ratio of a working condition is smaller than the ratio threshold, then the working condition is considered as an uncommon working condition.

Step 4 Generate new data samples for the uncommon working conditions; the balanced data sample sets are denoted as $\left\{\mathbf{S}_1^*, \mathbf{S}_2^*, ..., \mathbf{S}_{N_w}^*\right\}$.

3.3.2.3 Parameter identification based on EA-PSO

After sample labeling and balancing of the data samples, the unknown model parameters for each working condition can be identified separately. Parameter identification is essentially an optimization problem. The objective is to find a set of model parameters which fits best with the practical process dynamics. For the copper removal process, as its dynamics are nonlinear, an intelligent optimization algorithm is a feasible approach for solving the identification problem.

The particle swarm optimization (PSO) algorithm is a widely applied intelligent optimization algorithm [19,20]. To identify model parameters using the PSO algorithm, the position of a particle acts as the candidate of the unknown model parameters. In each iteration, by learning from the best particle of the swarm and its own best history experience, a particle updates its own position and velocity. After certain iterations, the algorithm converges to a potential optimum. The physical limits of the model parameters are treated as the range of particle positions. The advantage of PSO lies in its searching speed. However, it can easily get trapped in a local optimum. To overcome this drawback, an evolutionary algorithm (EA) is integrated in the PSO algorithm. In the searching process, the EA mutates some of the particles during the iteration, which increases the diversity of the particles and can help the trapped particles escape from the potential local optimum.

In the EA-PSO algorithm, the objective is to find a set of model parameters which minimizes the sum of errors between the model output and the real value of copper ion concentration of each data sample:

$$min J(\mathbf{p}_{\mathrm{mp}}) = \sum_{i=1}^{m} (\mathbf{C}_{\mathbf{Cu}^{2+}_{\mathrm{real}}} - \mathbf{C}_{\mathbf{Cu}^{2+}})^2, \tag{3.34}$$

where m is the number of data samples. The velocity and position of particles are updated as

$$\begin{aligned} v_i(k+1) &= wv_i(k) + c_1 R_1(pIbest_i(k) - x_i(k)) \\ &\quad + c_2 R_2(pGbest(k) - x_i(k))x_i(k+1) \\ &= x_i(k) + v_i(k+1), \end{aligned} \tag{3.35}$$

where k indicates the current iteration, $pIbest_i$ is the best position found by the ith particle, $pGbest$ is the best position found by the whole swarm, w is an inertial weight, c_1 and c_2 are constants, R_1 and R_2 are random numbers within the range [0, 1], and v_i and x_i are the velocity and position of the ith particle, respectively.

The detailed procedure of the EA-PSO algorithm is as follows:

Step 1 Initialize the position and velocity of N_{popul} particles according to the dimension and limits of the solution space. Define the threshold of fitness value as $T_{fitness}$:

$$X_i(0) = x_{ij}(0) \ (i = 1, ..., N_{\mathrm{popul}};\ j = 1, ..., m),$$
$$V_i(0) = v_{ij}(0) \ (i = 1, ..., N_{\mathrm{popul}};\ j = 1, ..., m).$$

Step 2 Let $k = 0$. Assign the initial fitness value of each particle to $pIbest_i(0)$, and the minimum fitness value of the swarm to $pGbest(0)$.

Step 3 Update the velocity and position of each particle.

Step 4 Calculate the fitness of all particles, and obtain $pIbest(k)$ and $pGbest(k)$.
Step 5 For a particle, if the fitness is larger than T_{fitness}, then mutate the particle according to EA. Let $k = k + 1$.
Step 6 If the algorithm reaches the maximum iteration number or the global best solution remains constant for several iterations, then stop the iteration and output the global optimum. If not, return to Step 2.

For different working conditions, the kinetic parameters are optimized through this procedure. The kinetic model for the ith condition can be expressed as $\mathbf{MODEL}(t, \mathbf{p}_K, \mathbf{p}_{\text{con}i})$.

In order to use the model online, a procedure named multicondition competitive-consecutive reaction system model (MCRSM) has been proposed, as shown in Fig. 3.5. The procedure consists of three parts, including a working condition recognition module, a multicondition parameter selector, and a CCRS model. First, the real-time industrial data are preprocessed and sent to the working condition recognition module, which estimates the type of current working condition. The working condition recognition module is trained using a multiclass SVM. After the type of current working condition is recognized, the corresponding model parameter set is selected and the kinetic model can be used to predict the outlet copper ion concentration.

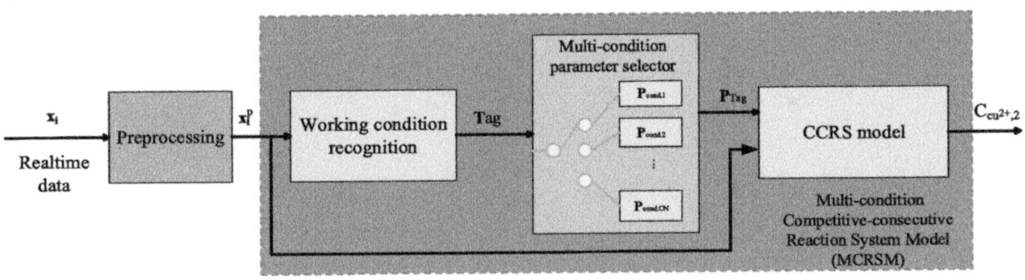

FIGURE 3.5 Multicondition competitive-consecutive reaction system model.

3.3.2.4 Results

Based on historical production data, the modeling and data analysis of the copper removal process were performed. One thousand groups of data were randomly selected as training samples. The data were classified into four working conditions using the proposed data sample labeling and classification method. The model parameters under each working condition were identified to establish the multicondition competitive-consecutive reaction system model. In order to verify the model accuracy, the production data of copper removal under three typical operating states were selected for testing, and 50 groups of "continuous-in-time" data samples were selected as test samples of each working condition. The fluctuation ranges of the process variables under the three typical operating states are shown in Table 3.2. Among these process parameters, the ranges of flow rate of underflow, solid content in the underflow, pH value, and tempera-

TABLE 3.2 Fluctuation ranges of the process variables under the three typical operating states.

Operating state	Steady state	Periodic fluctuation	Fluctuates with a trend
C_{Cu}^{Feed}	0.9–1.0	0.7–1.2	0.8–1.2
Q	295–305	180–260	200–250
Q_{uf}	29–31	28–30	29–30
g_{under}	1.4–1.5	1.4–1.5	1.4–1.5
g_{Zn}	120–145	60–156	70–150
pH	3.9–4.2	3.9–4.2	3.9–4.2
Temperature	62–66	62–66	62–66

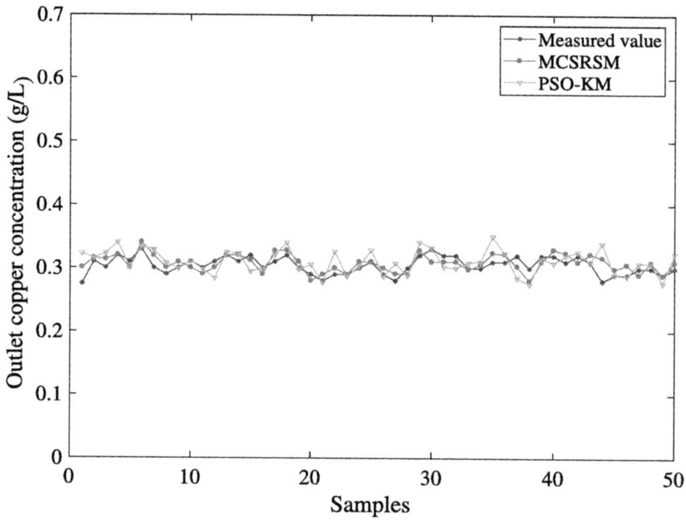

FIGURE 3.6 Comparison of MCRSM and PSO-KM under steady-state conditions.

ture are stable. While other parameters have smaller fluctuation ranges in steady state, they have larger variations in fluctuation states.

For comparison purposes, a PSO-based kinetic model (PSO-KM) was introduced in the experiment. The PSO-KM only identifies the model parameters using PSO. However, the working conditions were not divided. Both models were applied to the prediction of the outlet copper ion concentration. The prediction results are shown in Figs. 3.6 to 3.8. It can be observed that:

(i) When the copper removal process runs under steady-state conditions, MCRSM and PSO-KM exhibit similar performance in following the variation trend of copper ion concentration. The performance of MCRSM is slightly better than that of the PSO-KM, as shown in Fig. 3.6.

(ii) When the operation state fluctuates, the overall prediction performance of MCRSM is better than that of PSO-KM, as shown in Figs. 3.7 and 3.8.

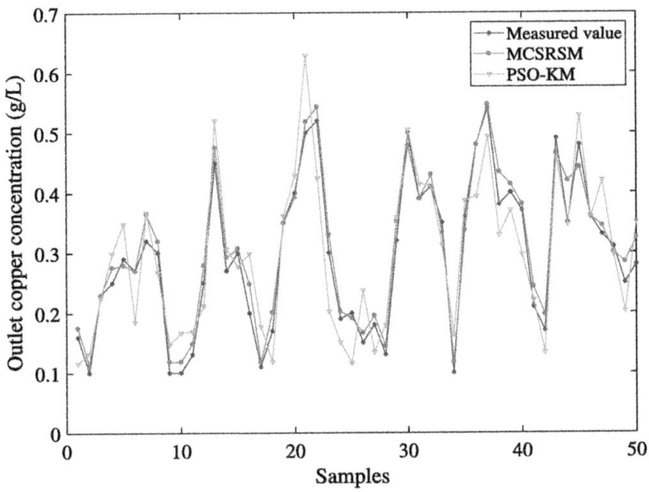

FIGURE 3.7 Comparison of MCRSM and PSO-KM under periodic fluctuations.

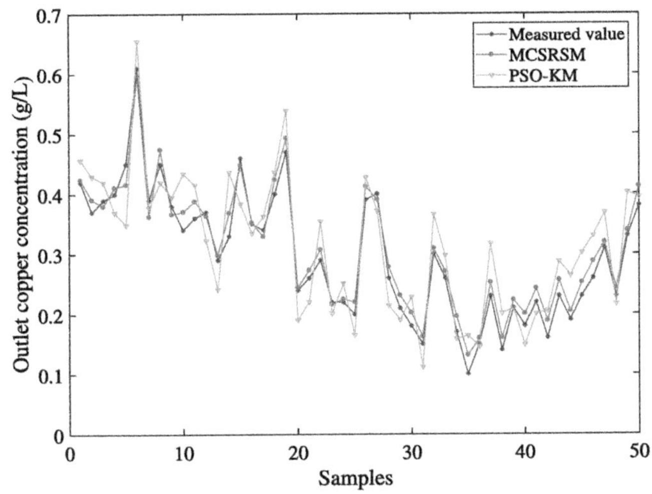

FIGURE 3.8 Comparison of MCRSM and PSO-KM under fluctuations with a trend.

(iii) When the operating state fluctuates periodically, the accuracy of MCRSM is higher than that of PSO-KM. When the operating state fluctuates with a trend, the accuracy of PSO-KM is close to that of MCRSM. However, the error of MCRSM is more concentrated than that of PSO-KM.

Considering a long time range, the copper removal process can switch among various different working conditions. As a result, the copper ion concentration fluctuates greatly. PSO-KM regards the copper removal process as a single steady-state process. It cannot describe the transition between multiple

TABLE 3.3 Identification results of model parameters under different working conditions.

Working condition	$K_D \times 10^{-5}$	$k_G \times 10^{-5}$	$D_e \times 10^{-7}$	$k_{surf} \times 10^{-5}$
I	4.97 ± 0.08	3.21 ± 0.09	3.17 ± 0.13	4.52 ± 0.11
II	5.13 ± 0.11	3.42 ± 0.12	3.27 ± 0.10	4.71 ± 0.13
III	5.38 ± 0.13	3.61 ± 0.14	3.44 ± 0.12	5.02 ± 0.18
IV	4.61 ± 0.14	3.09 ± 0.10	3.01 ± 0.09	4.32 ± 0.08

TABLE 3.4 Numbers of recognized working conditions under different operating states.

Operating state	Working condition I	Working condition II	Working condition III	Working condition IV
Steady state	41	9	0	0
Periodic fluctuation	18	14	16	12
Fluctuation with a trend	20	8	2	10

working conditions. Therefore, it is difficult to track the variation trend of copper ion concentration accurately. However, the MCRSM can detect most transitions of the working conditions by perceiving the variation of data features. This ensures the accuracy of the model under fluctuating working conditions. Table 3.3 shows the model parameters of the four working conditions, which are identified by the MCRSM. Under different working conditions, the model parameters are different. As a result, the consumption rate of copper ion concentration changes with working conditions. Table 3.4 lists the recognition result of the working conditions using MCRSM. When the process is operating under steady-state conditions, the switching frequency of working conditions is low. The number of recognized working conditions is small. When the operation fluctuates, the working condition is prone to change, and the duration of each working condition is short. The number of recognized working conditions is large. This indicates that the working condition recognition result of MCRSM agrees with the practical situation of the copper removal process. Therefore, the MCRSM can track and better describe the dynamics of the practical copper removal process.

In order to evaluate the model accuracy quantitatively, four indicators are introduced:

- mean absolute error (MAS),
- root mean square error (RMSE),
- maximum absolute error (MAE),
- accuracy rate (accuracy probability [ΛP]),

where AP is designed according to the production requirement. It measures the similarity between the predicted value and the measured value,

$$A_c = P(Y_e(i) \in [Y_m(i) - \eta, Y_m(i) + \eta]), \tag{3.36}$$

where

$$Y_e(i) = Y_c(i) - Y_m(i), \tag{3.37}$$

$$\eta = \begin{cases} 0.02, & 0 < Y_m(i) \leqslant 0.15, \\ 0.03, & 0.15 < Y_m(i) \leqslant 0.45, \\ 0.04, & 0.45 \leqslant Y_m(i), \end{cases}$$

where $Y_c(i)$ is the model output, $Y_m(i)$ is the measured value, $Y_e(i)$ is the error between $Y_c(i)$ and $Y_m(i)$, and η is the precision threshold.

TABLE 3.5 Statistical results of MCRSM and PSO-KM.

Operating state	Method	MAE (g/L)	RMSE	MAXABS (g/L)	AP
Steady state	MCRSM	0.010	0.019	0.036	0.96
	PSO-KM	0.013	0.020	0.038	0.91
Periodic fluctuation	MCRSM	0.021	0.023	0.078	0.85
	PSO-KM	0.029	0.026	0.078	0.80
Fluctuation with a trend	MCRSM	0.017	0.014	0.041	0.90
	PSO-KM	0.021	0.019	0.058	0.86

The statistical analysis result of the model accuracy under the three working conditions is shown in Table 3.5. When the copper removal process is under steady-state conditions, the performance of PSO-KM is satisfying. It can effectively predict the change of outlet copper ion concentration. However, when the operating state fluctuates with a clear trend, the AP of MCRSM is 5% higher than that of PSO-KM. The other indicators of MCRSM are also better than PSO-KM. When the running state fluctuates periodically, the MAE and RMSE of MCRSM are significantly reduced compared with PSO-KM. In addition, the AP of MCRSM is 5% higher than that of PSO-KM. This illustrates that the advantage of MCRSM is more remarkable when the operating state fluctuates.

References

[1] H. Zhang, Y. Li, J. Wang, X. Hong, The influence of nickel ions on the long period electrowinning of zinc from sulfate electrolytes, Hydrometallurgy 99 (1–2) (2009) 127–130.

[2] A.M. Bond, 200 years of practical electroanalytical chemistry: past, present and future directions illustrated by reference to the online, on-stream, and offline determination of trace metals in zinc plant electrolyte by voltammetric and potentiometric techniques, Analytica Chimica Acta 400 (1–3) (1999) 333–379.

[3] K. Laatikainen, M. Lahtinen, M. Laatikainen, E. Paatero, Copper removal by chelating adsorption in solution purification of hydrometallurgical zinc production, Hydrometallurgy 104 (1) (2010) 14–19.

[4] W. Djoudi, F. Aissani-Benissad, S. Bourouina-Bacha, Optimization of copper cementation process by iron using central composite design experiments, Chemical Engineering Journal 133 (1–3) (2007) 1–6.

[5] F. Gros, S. Baup, M. Aurousseau, Intensified recovery of copper in solution: Cementation onto iron in fixed or fluidized bed under electromagnetic field, Chemical Engineering and Processing: Process Intensification 47 (3) (2008) 295–302.
[6] S. Nosier, Y. Alhamed, H. Alturaif, Enhancement of copper cementation using ceramic suspended solids under single phase flow, Separation and Purification Technology 52 (3) (2007) 454–460.
[7] C. Noubactep, Elemental metals for environmental remediation: learning from cementation process, Journal of Hazardous Materials 181 (1–3) (2010) 1170–1174.
[8] J. Näsi, Statistical analysis of cobalt removal from zinc electrolyte using the arsenic-activated process, Hydrometallurgy 73 (1–2) (2004) 123–132.
[9] L. Lehtinen, M. Lahtinen, M. Jyrälä, M. Vuokko, Method for the removal of copper from a zinc sulphate solution, 2010, US Patent 7,682,581.
[10] I. Yahiaoui, F. Aissani-Benissad, Experimental design for copper cementation process in fixed bed reactor using two-level factorial design, Arabian Journal of Chemistry 3 (3) (2010) 187–190.
[11] A. Dib, L. Makhloufi, Cementation treatment of copper in wastewater: mass transfer in a fixed bed of iron spheres, Chemical Engineering and Processing: Process Intensification 43 (10) (2004) 1265–1273.
[12] B. Dönmez, F. Sevim, H. Saraç, A kinetic study of the cementation of copper from sulphate solutions onto a rotating aluminum disc, Hydrometallurgy 53 (2) (1999) 145–154.
[13] B. Zhang, C. Yang, H. Zhu, Y. Li, W. Gui, Kinetic modeling and parameter estimation for competing reactions in copper removal process from zinc sulfate solution, Industrial & Engineering Chemistry Research 52 (48) (2013) 17074–17086.
[14] J. Sedzimir, Precipitation of metals by metals (cementation)—kinetics, equilibria, Hydrometallurgy 64 (3) (2002) 161–167.
[15] T. Knorr, M. Kaiser, F. Glenk, B.J. Etzold, Shrinking core like fluid solid reactions—a dispersion model accounting for fluid phase volume change and solid phase particle size distributions, Chemical Engineering Science 69 (1) (2012) 492–502.
[16] C. Wen, Noncatalytic heterogeneous solid-fluid reaction models, Industrial and Engineering Chemistry 60 (9) (1968) 34–54.
[17] Y. Li, W. Gui, K.L. Teo, H. Zhu, Q. Chai, Optimal control for zinc solution purification based on interacting CSTR models, Journal of Process Control 22 (10) (2012) 1878–1889.
[18] H. He, E. Garcia, Learning from imbalanced data, IEEE Transactions on Knowledge and Data Engineering 21 (9) (2009) 1263–1284.
[19] J. Prasad, T. Souradeep, Cosmological parameter estimation using particle swarm optimization (PSO), Physical Review D 85 (12) (2011) 123008.
[20] M. Schwaab, E.C. Biscaia Jr., J.L. Monteiro, J.C. Pinto, Nonlinear parameter estimation through particle swarm optimization, Chemical Engineering Science 63 (6) (2008) 1542–1552.

Chapter 4

Additive requirement ratio estimation using trend distribution features

Contents

4.1 Definition of additive requirement ratio

In the copper removal process, zinc powder is used as additive to replace the impurity copper ions [1]. According to the chemometrics of copper removal reactions, the dosage of additive is crucial for copper removal. On the one hand, the additive dosage directly affects the reaction impetus and impurity removal performance of each reactor. On the other hand, the additive is not fully utilized in impurity removal as the practical production conditions are not ideal, and the utilization rate of the additive varies with the inlet and reaction conditions [2]. However, due to the complex reaction mechanism and time-varying working conditions, it is difficult for human operators to estimate the additive utilization rate. As a result, the precise determination of additive dosage is challenging [3]. Hence, in this chapter, an estimation framework for predicting the additive utilization ratio is designed to eliminate this drawback [4].

To analyze the estimation problem, the following variables are defined:

- $\mathbf{I}_i$: the variable vector of the inlet conditions of the ith reactor ($i = 1, 2, \cdots, n$), which consists of the impurity concentrations, the flow rate of the inlet solution, etc.,
- A_i: the amount of additive added to the ith reactor ($i = 1, 2, \cdots, n$),
- $\mathbf{P}_i$: the variable vector of the reaction conditions of the ith reactor ($i = 1, 2, \cdots, n$), including the temperature, pH of the solution, volume of the reactor, etc.,
- C_i^{out}: the outlet impurity concentration of the ith reactor ($i = 1, 2, \cdots, n$).

Modeling, Optimization, and Control of Zinc Hydrometallurgical Purification Process
https://doi.org/10.1016/B978-0-12-819592-5.00014-4

According to the chemometrics, the ideal additive amount added in the ith reactor at time step k can be formulated as follows:

$$A_i^M(k) = f_i^M(\mathbf{I}_i(k), \mathbf{P}_i(k), C_i^{goal}), \tag{4.1}$$

where $A_i^M(k)$ is the ideal additive amount of the ith reactor at time step k, $f_i^M(\bullet)$ represents the additive model of the ith reactor, C_i^{goal} denotes the target outlet impurity concentration of the ith reactor, and $\mathbf{I}_i(k)$ and $\mathbf{P}_i(k)$ denote the values of the inlet variables (flow rate, inlet concentrations, etc.) and the reaction condition variables (pH value, temperatures, etc.) of the ith reactor during the time interval of interest, respectively.

It is difficult to control the outlet impurity concentration precisely at its target value due to the frequent fluctuations of inlet and reaction conditions. The main reason behind this drawback is the inequality between the actually required additive dosage and the ideal additive dosage. Considering this problem, a concept named additive requirement ratio (ARR) and its estimation framework are proposed in this chapter.

As shown in Eq. (4.2), ARR is defined as the ratio between the actually required additive dosage to achieve the expected impurity removal performance and the ideal additive dosage calculated by Eq. (4.1):

$$E_i^A(k) = A_{i,real}^{goal}(k) / f_i^M(\mathbf{I}_i(k), \mathbf{P}_i(k), C_i^{goal}), \tag{4.2}$$

where $A_{real}^{goal}(k)$ indicates the actually required additive dosage to reach the desired impurity removal performance. In Eq. (4.2), $f_i^M(\mathbf{I}_i(k), \mathbf{P}_i(k), C_i^{goal})$ can be obtained by the theoretical additive dosage model. However, the exact value of $A_{real}^{goal}(k)$ cannot be calculated online.

To solve the above problem, it is assumed that the ARR is affected only by the inlet and reaction conditions rather than the outlet impurity concentration. In other words, under the same inlet and reaction conditions, the ARRs are the same:

$$E_i^A(k) = \frac{A_{i,real}^{goal}(k)}{f_i^M(\mathbf{I}_i(k), \mathbf{P}_i(k), C_i^{goal})} = \frac{A_{i,real}^{free}(k)}{f_i^M(\mathbf{I}_i(k), \mathbf{P}_i(k), C_{i,outlet}^{free})}, \tag{4.3}$$

where $C_{i,outlet}^{free}$ is the target outlet impurity concentration of the ith reactor with an arbitrary value and $A_{i,real}^{free}(k)$ is the actually required additive dosage to reduce the impurity concentration to the target value. Hence, ARR can be reformulated using the following equation:

$$E_i^A(k) = \frac{A_i(k)}{f_i^M(\mathbf{I}_i(k), \mathbf{P}_i(k), C_i^{out}(k+1))}, \tag{4.4}$$

where $A_i(k)$ is the average actual additive consumption during the interval between time steps k and $k+1$ and $C_i^{out}(k+1)$ is the impurity concentration at the outlet of the ith reactor at time step $k+1$.

In most cases, zinc powder is added to each reactor to replace the impurity copper ions [5]. However, there exist some situations in which zinc powder is not added to each reactor. Considering the latter more general case, if we assume zinc powder is only added to the ith reactor, rather than the (i)th reactor to the $(i+l)$th reactor, then the ith to the $(i+l)$th reactors can be viewed as one reactor. The volume and residence time of the reactor equal the sums of those of each reactor. Therefore, the formulation of ARR can be expressed as

$$E_{i+l}^{A}(k) = \frac{A_i(k)}{f_i^M(\mathbf{I}_i(k), \mathbf{P}_i(k), C_{i+l}^{out}(k+1))}, \tag{4.5}$$

where $E_{i+l}^{A}(k)$ and $A_i(k)$ are the ARR and the average additive consumption of the ith to $(i+l)$th reactors during the interval between time steps k and $k+1$, $\mathbf{I}_i(k)$ is the average of the inlet conditions of the ith reactor between time steps k and $k+1$, $\mathbf{P}_i(k)$ is the average of the reaction conditions of the $(l+1)$ reactors between time steps k and $k+1$, and $C_{i+l}^{out}(k+1)$ is the outlet impurity concentration of the $(i+l)$th reactor at time step $k+1$.

By using Eqs. (4.4) and (4.5), the historical value of ARR can be calculated. In practice, however, the value of ARR varies under different working conditions. Under time-varying working conditions, the ARR of the last time interval does not equal the current ARR [6]. Therefore, the ARR should be predicted to provide useful information for making control decisions.

4.2 Case-based prediction with trend distribution features for ARR

According to the reaction kinetics and chemometrics, the value of ARR is determined by the numbers and types of chemical reactions in which the additive participates. Therefore, the inlet and reaction conditions (e.g., concentrations of reactants and products, pH, temperature, ORP) can affect ARR. In the copper removal process, the volume of a chemical reactor is large. The residence time of the leaching solution and additive in each reactor is long. The variations of process variables during the residence time, which form the reaction atmosphere, can affect ARR. Therefore, it is the trend of process variables rather than their values at a moment that determines the value of ARR. Correspondingly, the trend of process variables should be used as input to reconstruct the ARR. A case-based prediction framework with trend distribution feature [7] (CBP-TDF) is shown in Fig. 4.1. The proposed CBP-TDF consists of three parts. In the framework, the trends of process variables are extracted and classified first. Then, their distribution features are obtained. Finally, the ARR is predicted using an improved CBR algorithm with the process variables and their distribution features as inputs.

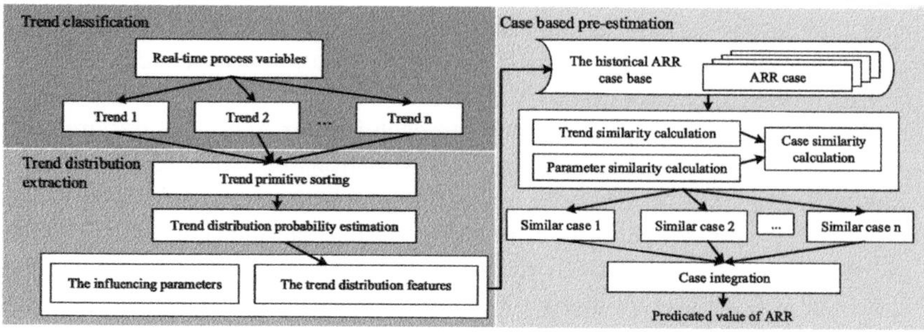

FIGURE 4.1 The framework for the estimation of the ARR.

4.2.1 Variation trend extraction and classification

In practice, the measurements can be classified into two categories, including online measured process variables and discrete-delayed measured KPI or copper ion concentration. The process variables change constantly during the offline sampling period of copper ion concentration. The variation trends of these process variables indicate the underlying change of copper ion concentration. As the copper ion concentration cannot be obtained online, the variation trends of these process variables are used to reveal the potential change of ARR.

TABLE 4.1 Physical meaning of the first and second derivatives when taking different values.

Type	Positive (+)	Negative (−)	Zero (0)
First derivative	Rising	Falling	Stable
Second derivative	Change rate is increasing	Change rate is decreasing	Change rate is stable

Qualitative shape analysis (QSA) is one of the most common trend analysis methods [8]. In this method, the primitives, which are a set of qualitative labels, are used to describe the variation of variables of interest. Firstly, the first and second derivatives of process variables are calculated. The physical meaning of the first and second derivatives are listed in Table 4.1. As indicated by Table 4.1, the trend of process variables can be divided into seven primitives according to the values of their first and second derivatives, i.e., positive (+), negative(−), or zero(0), as shown in Fig. 4.2. For example, if both the first and second derivatives are negative, there is a nonlinear declining trend at this data point, which is denoted as (−, −) or G. Through the combining of the primitives, the process trend could be represented hierarchically.

By using the QSA approach [8], the trend of a process variable, which is extracted by its qualitative shapes, can be described by a series of primitives. However, the variation quantity of the variables, which can reveal the state of

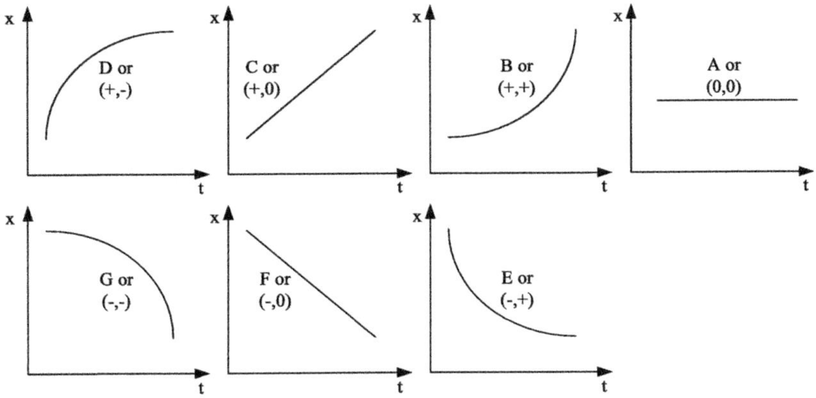

FIGURE 4.2 A sketch of primitives (A, B, C, D, E, F, and G are labels for the primitives).

the removal reaction, is not considered in the QSA method. For example, the amount of impurities removed during the reaction time is related to the change rate of ORP. The decision of additive dosage may be not appropriate and the precision of ARR estimation could be reduced when the change rate of ORP is neglected [9]. Therefore, in the proposed CBP-TDF approach, the performance of trend classification is improved by also taking the variation quantity into consideration.

4.2.1.1 Smoothing and normalization of process variables

Suppose $\mathbf{x} = \{x_1, x_2, ..., x_m\}$ is the time series of process variables with length m in the copper removal process. Before extracting the trend of $\mathbf{x}$, it should be smoothed and normalized. On the one hand, through smoothing of the real-time process variables, some high-frequency noises which could lead to misclassifications among the trends can be decreased [10]. On the other hand, the min/max normalization approach can reduce the influences of different dimensions, which allows the comparison of variables with different orders of magnitude. During the normalization of $\mathbf{x}$, the minima and maxima are chosen from the historical data set instead of the current interval. Therefore, a normalized variable vector $\mathbf{x}' = \left\{x'_1, x'_2, ..., x'_m\right\}$ with a transformed range of [0, 1] is obtained.

4.2.1.2 Differentiation of process variables and setting primitive thresholds

For $\mathbf{x}$, its first derivative vector $\mathbf{x}_d = \{d_1, d_2, ..., d_{m-1}\}$ is produced from the normalized variables by the discrete approximation of differentiation:

$$d_i = x'_i - x'_{i-1}, \quad i > 0. \tag{4.6}$$

The second derivative vector $\mathbf{x}_\delta = \{\delta_1, \delta_2, ..., \delta_{m-2}\}$ can be obtained by differentiating $\mathbf{x}_d = \{d_1, d_2, ..., d_{m-1}\}$ in the same way:

$$\delta_i = d_i' - d_{i-1}', \quad i > 0. \tag{4.7}$$

The range of first derivatives is $[-1, 1]$, while the range of second derivatives is $[-2, 2]$. In order to classify the change rate of real-time process variables, four thresholds of the first derivative vector are defined based on the historical data set:

$$th_1^- = \mathrm{Q}_{1/4}(\mathbf{x}_d^-), \tag{4.8}$$
$$th_2^- = \mathrm{Me}(\mathbf{x}_d^-), \tag{4.9}$$
$$th_1^+ = \mathrm{Me}(\mathbf{x}_d^+), \tag{4.10}$$
$$th_2^+ = \mathrm{Q}_{3/4}(\mathbf{x}_d^+), \tag{4.11}$$

where $\mathbf{x}_d^- = \{d_i | d_i < 0, d_i \epsilon D\}$, $\mathbf{x}_d^+ = \{d_i | d_i > 0, d_i \epsilon D\}$, D is the derivative calculated by all historical data instead of the limited data sampled from the current interval, Me denotes the medium, $-1 < th_1^- < th_2^- < 0 < th_1^+ < th_2^+ < 1$, and $\mathrm{Q}_{1/4}$ and $\mathrm{Q}_{3/4}$ represent the first and third quartiles, respectively. Through the definition of thresholds, the trends of process variables can be divided into seven types, which are labeled according to the magnitude of change (increase, stable, and decrease) [10]. All seven types of trends are described in Table 4.2. Similarly, in each type of trends, they are comprised of three classes according to their second derivative values, which are shown in Table 4.3.

TABLE 4.2 Variation trend classification using first derivatives.

Label	Range	Symbol
Rapid decrease	$d_i \in [-1, th_1^-]$	D(H, ∗)
Moderate decrease	$d_i \in [th_1^-, th_2^-]$	D(M, ∗)
Slight decrease	$d_i \in [th_2^-, 0]$	D(L, ∗)
Steady	$d_i = 0$	S(0, 0)
Slight increase	$d_i \in (0, th_1^+]$	I(L, ∗)
Moderate increase	$d_i \in (th_1^+, th_2^+]$	I(M, ∗)
Rapid increase	$d_i \in (th_2^+, 1]$	I(H, ∗)

4.2.1.3 *Identifying trends*

Based on the above definition, 19 qualitative primitives can be generated, including 9 types of increasing trend, 1 type of stable trend, and 9 types of decreasing trend. As shown in Fig. 4.3, ∗(H, ∗), ∗(M, ∗), ∗(L, ∗), and (0, ∗) denote the absolute values of the first derivative of a variable which are high, moderate, low,

TABLE 4.3 Variation trend classification using second derivatives.

Label	Range	Symbol
Concave shape	$\delta_i \in [-2, 0)$	D(∗, −) or I(∗, −)
Linear shape	$\delta_i = 0$	D(∗, 0) or I(∗, 0)
Convex shape	$\delta_i \in (0, 2]$	D(∗, +) or I(∗, +)

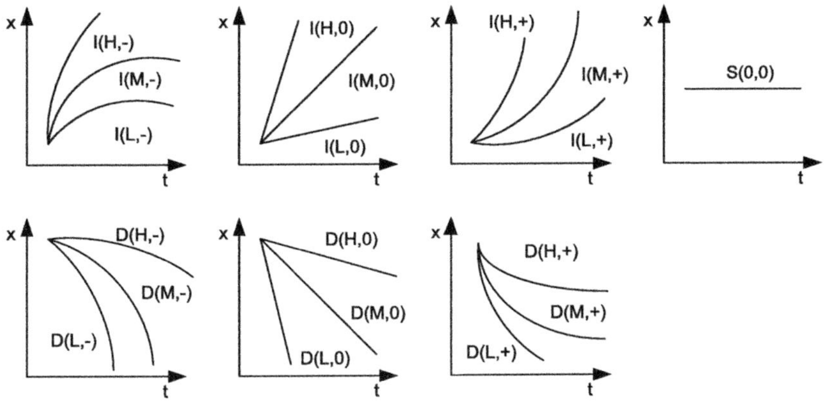

FIGURE 4.3 The primitives in the CBP-TDF method.

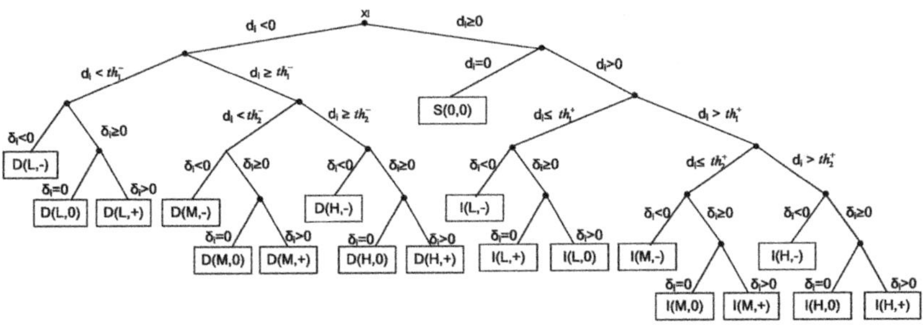

FIGURE 4.4 Decision tree for trend classification: x_i denotes the ith variable and d_i and δ_i denote the first and second derivatives of x_i, respectively.

and zero, respectively, ∗(∗, +), ∗(∗, −), and ∗(∗, 0) denote the positive, negative, and zero second derivative of the variables, respectively, and D, I, and S denote the decrease, increase, and stable trend, respectively. The decision tree for classifying the qualitative primitives is shown in Fig. 4.4.

4.2.2 Extracting trend distribution features

In practical process, the state of a variable is identified as stable if its trend is labeled as steady, and unstable if its trend is labeled as rapid increase or decrease. The distribution of variation trends during a time interval reveals the state of the corresponding process variable. The trends under similar operating conditions are supposed to hold similar distributions even if they are sampled from different time intervals. Therefore, if the features of the trend distributions are extracted, the similarities of different process conditions can be evaluated by comparing the trend distribution features. In the rest of this section, the steps in extracting the trend distribution features are introduced, including sorting of qualitative primitives and distribution feature extraction based on the probability density function.

4.2.2.1 *Sorting the qualitative primitives*

Sorting the qualitative primitives is the first step in obtaining the characteristics of the trend distribution. For better extraction of the trend distribution features, an appropriate logic sort order is required. Then, a set of sorting rules based on the change rates of qualitative primitives can be formulated. In the sorting rules, the trend labeled steady $S(0, 0)$ is set as the origin, which can be viewed as an equilibrium point of a variable. Downward primitives are labeled as negative numbers, representing decrease of a variable. Upward primitives are set as positive numbers, representing increase of a variable.

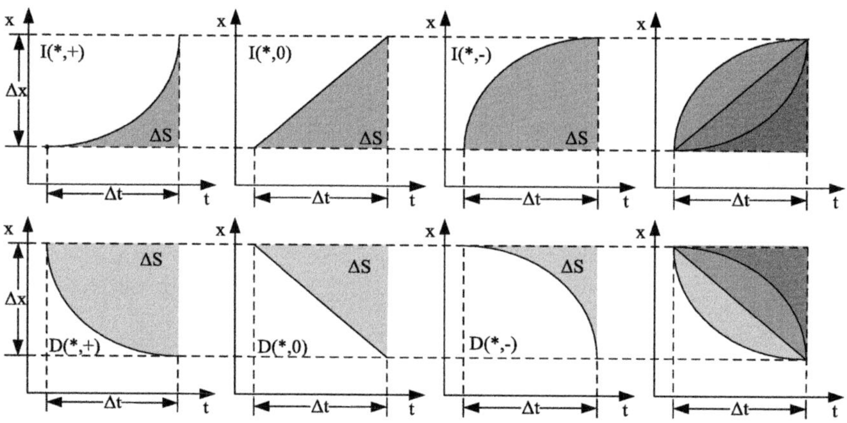

FIGURE 4.5 Cumulative variations of a primitive in unit time.

The cumulative variations of primitives with different shapes are shown in Fig. 4.5, where the shade area is the cumulative decrement or cumulative increment of a primitive. As indicated in Fig. 4.5, in unit time, the cumulative increment/decrement of a variable with the same increment/decrement (Δx) varies with the shapes of variation. If a variable increases, the concave shape has the smallest cumulative increment, while the convex shape has the largest cumulative increment. On the contrary, if a variable decreases, the concave shape has the smallest cumulative decrement, while the convex shape has the largest cumulative decrement. When the variation trends have the same shape, the value of cumulative increment or decrement depends on the increase or decrease rate of a variable. When the trend label is less than zero, then the smaller the label is, the larger the cumulative amount of the variable decreased per unit time is. If the trend label is greater than zero, then the greater the label is, the larger the cumulative amount of the variable increased per unit time is. This method realizes the quantitative description of trend distribution of a primitive. The sorting of the 19 primitives is shown in Table 4.4.

TABLE 4.4 Range divisions and symbol assignments for the second derivatives.

Primitive	Label	Primitive	Label	Primitive	Label
D(H, +)	−9	D(L, 0)	−2	I(M, 0)	5
D(H, 0)	−8	D(L, −)	−1	I(M, −)	6
D(H, −)	−7	S(0, 0)	0	I(H, +)	7
D(M, +)	−6	I(L, +)	1	I(H, 0)	8
D(M, 0)	−5	I(L, 0)	2	I(H, −)	9
D(M, −)	−4	I(L, −)	3		
D(L, +)	−3	I(M, +)	4		

4.2.2.2 *Estimating the trend distribution probability*

The shape of the trend distribution is an external reflection of the internal state of variables within an interval. Therefore, an adequate distribution function is required to describe the trend distribution accurately. The trend distribution of an industrial variable over a short time interval is not exactly the same as the conventional distribution models (e.g., Weibull, Rayleigh, Gaussian, etc.). Using the conventional distribution models to estimate the trend distribution may generate large errors. Thus, as a nonparametric method for estimating the probability density, the kernel density estimation approach is used to evaluate the statistical properties of trend distribution [11].

If we denote the frequencies of the 19 qualitative primitives as a vector $\mathbf{s} = \{s_{-9}, \cdots, s_{-1}, s_0, s_1, \cdots, s_9\}$, then the probability density function of the primitives is estimated through a kernel density function:

$$P(s) = \frac{1}{19h} \sum_{i=-9}^{9} K(\frac{s - s_i}{h}), \tag{4.12}$$

where h denotes the bandwidth which could be selected by the plugin rule procedure [11], $K(\bullet)$ is the kernel function, and $P(s)$ is the trend distribution feature describing the process state.

4.2.3 Case-based prediction with a trend distribution feature

The principle of case-based prediction (CBP) is reconstructing ARR using cases retrieved from a database with high similarity with the current process [12][13].

The similarities between cases, which consist of process variables, are calculated using a distance function. However, the conventional CBP is not appropriate for predicting ARR when the cases consist of trend distribution features. The reason is that the trend distribution features consist of a series of probability density functions. The similarity between these functions cannot be calculated by the distance function adopted in the conventional CBP. Therefore, it is necessary to develop a case-based prediction strategy [14] which can calculate the similarity between both trend distribution features, which is measured through the Kullback–Leibler (KL) divergence [15], and process variables.

4.2.3.1 Similarity measurements for the trend distributions and industrial variables

Denote $\mathbf{C}_k = \{v_{k1}, \cdots, v_{km}, P_{k1}(s), \cdots, P_{kb}(s), E_k\}$ as the kth case in the case database, where v_{kj} is the value of the jth process variable ($j = 1, \cdots, m$) in the case, $P_{kj}(s)$ is the jth trend distribution feature ($j = 1, \cdots, b$), and E_k is the value of ARR. Denote $\mathbf{I}_{input} = \{v_{input\cdot 1}, \cdots, v_{input\cdot m}, P_{input\cdot m}(s), \cdots, P_{input\cdot b}(s)\}$ as the input vector of the ARR prediction algorithm, where $v_{input\cdot j}$ ($j = 1, \cdots, m$) and $P_{input\cdot i}(s)$ ($i = 1, \cdots, b$) are the process variables and the trend distribution features, respectively. Consequently, the similarity between the jth trend distribution feature of the current input vector and the jth trend distribution feature of the kth case in the database is

$$S_{TD}(P_{input\cdot j}, P_{kj}) = \sum_{s} P_{input\cdot j}(s) \log \frac{P_{input\cdot j}(s)}{P_{kj}(s)}. \tag{4.13}$$

The total similarity between all the trend distribution features of the current input vector and the kth case in the database is

$$S_{TD}(\mathbf{P_{input}}, \mathbf{P_k}) = \sum_{j=1}^{b} \alpha_j S_{TD}(P_{input\cdot j} \parallel P_{kj}), \tag{4.14}$$

where α_j is the weight of the jth trend distribution feature and $\sum_{j=1}^{b} \alpha_j = 1$.

The similarity between the process variables of the kth case $\mathbf{C}_k$ and the input vector I, which is expressed as $D(\mathbf{i}, \mathbf{v}_k)$, is calculated using the Euclidean distance function:

$$S_{PV}(\mathbf{v_{input}}, \mathbf{v_k}) = \sum_{j=1}^{m} w_j |i_j - v_{kj}|, \tag{4.15}$$

where w_j is the weight of the jth process variable and $\sum_{j=1}^{m} w_j = 1$.

4.2.3.2 Prediction of ARR

The similarity between the current input vector I and the kth case $\mathbf{C}_k$ is a combination of $S_{TD}(\mathbf{P_{input}}, \mathbf{P_k})$ and $S_{PV}(\mathbf{v_{input}}, \mathbf{v_k})$, which is formulated as

$$S_{total}(\mathbf{I}, \mathbf{C}_k) = \lambda S_{TD} S_{TD}(\mathbf{P_{input}}, \mathbf{P_k}) + (1-\lambda) S_{PV}(\mathbf{v_{input}}, \mathbf{v_k}), \tag{4.16}$$

where λ is the similarity weight.

A similarity threshold (th_{sim}) is used to screen the cases. The ARR is estimated as a weighted average of the ARRs of the selected cases:

$$E^P = \beta_i \sum_{i=1}^{Num} E_i^S, \tag{4.17}$$

where E^P is the estimated ARR, Num is the number of selected cases, and E_i^S is the ARR of the ith selected case. The weight β_i is the similarity ratio between the ith selected case and all the selected cases:

$$\beta_i = \frac{S_{total}(\mathbf{I}, \mathbf{C}_i)}{\sum_{i=1}^{Num} S_{total}(\mathbf{I}, \mathbf{C}_i)}. \tag{4.18}$$

It can be observed from the formulation that the performance of the TDF-CBP approach depends on the selection of weights ($\alpha_1 \sim \alpha_b,\ w_1 \sim w_m,$ and λ), which can be optimized by minimizing the differences between the estimated ARR and the actual ARR:

$$\begin{aligned}
&\min \quad J = \sqrt{\frac{\sum_{i=1}^{N_{opt}} [E_{actual\cdot i} - f_{TDF-CBP}(\mathbf{I}_i, \alpha_1, \cdots, \alpha_b; w_1, \cdots, w_m; \lambda)]^2}{N_{opt}}}, \\
&s.t. \quad 0 \le \alpha_j \le 1, j = 1, \cdots, b, \\
&0 \le w_j \le 1, j = 1, \ldots, m, \\
&0 \le \lambda \le 1, \\
&\sum_{j=1}^{b} \alpha_j = 1, \\
&\sum_{j=1}^{m} w_j = 1,
\end{aligned} \tag{4.19}$$

where $f_{TDF-CBP}$ is the estimated ARR using the TDF-CBP method, $\mathbf{I}_i$ ($i = 1, \cdots, N_{opt}$) is the ith input of the function, and $\mathbf{E}_{actual\cdot i}$ is the actual ARR corresponding to the ith input.

4.3 Results

To test the performance of the ARR estimation framework, a simulation is conducted using real industrial data covering two representative operating conditions. In the first type of operating condition (operating condition I), the process is relatively stable with slight oscillations in the ARR. In the second type of operating condition (operating condition II), the ARR varies dramatically due to the fluctuating inlet and reaction conditions. Under each operating condition, two control approaches determining additive dosage are applied. In the first control approach, additive dosage is calculated online using an additive model (AMC) which is built based on the stoichiometric relations in the copper removal reactions:

$$M_{AM} = (1 - 0.5\varsigma)(I_{Cu} - G_{Cu})F\Gamma, \tag{4.20}$$

where M_{AM} denotes the additive dosage, ς is a reaction ratio coefficient, I_{Cu} and G_{Cu} denote the inlet and target copper ion concentrations, respectively, F is flow rate of the inlet solution, and Γ is the ratio between the molar masses of zinc and copper. Therefore, considering a single reactor, the ARR of the AMC approach at time step k is

$$E_{AM}(k) = \frac{\bar{M}_{Zn}}{(1 - 0.5\varsigma)\left[I_{Cu}(k) - C_{Cu}^{out}(k+1)\right]\bar{F}\Gamma}, \tag{4.21}$$

where $C_{Cu}^{out}(k+1)$ is the outlet copper ion concentration at time step $k+1$, $G_{Cu}(k)$ is the inlet copper ion concentration at time step k, and $\bar{M}_{Zn}$ and $\bar{F}$ are the average mass flows of additive and average inlet flow rate between time steps k and $k+1$, respectively. In the second control approach, additive dosage is obtained using model predictive control (MPC):

$$M_{MPC} = f_{MPC}(I_{Cu}, G_{Cu}, F, T, e_{ORP}, e_{pH}), \tag{4.22}$$

where M_{MPC} denotes the additive dosage, T is temperature, e_{ORP} is ORP, and e_{pH} is the pH of the leaching solution. Then, similarly, the ARR of the MPC approach at time step k is

$$E_{MPC}(k) = \frac{\bar{M}_{MPC}}{f_{MPC}(I_{Cu}(k), C_{Cu}^{out}(k+1), F(k), T(k), e_{ORP}(k), e_{pH}(k))}, \tag{4.23}$$

where $\bar{M}_{MPC}$ is the average mass flow of additive between time steps k and $k+1$.

The ARRs under the four situations (two operating conditions, two different control approaches for each operating condition) are estimated using the proposed CBP-TDF approach. For comparison, the ARR is also predicted using

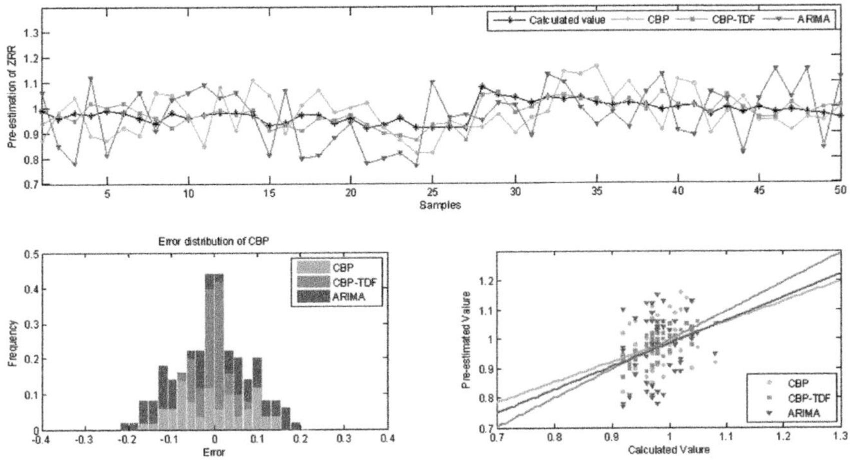

FIGURE 4.6 ARR estimation results when using the AMC approach under operating condition I.

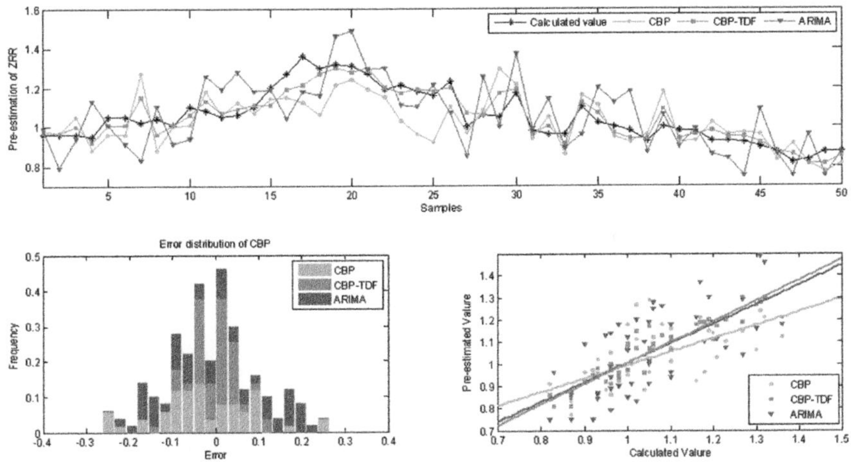

FIGURE 4.7 ARR estimation results when using the AMC approach under operating condition II.

conventional CBR and ARIMA algorithms. Four indices are selected to evaluate the estimation performance, including the mean absolute error (MAE), the deviation range (DR), the root mean square error (RMSE), and the p-value.

For the AMC approach, the estimation results of ARR under operating condition I using different approaches are shown in Fig. 4.6. It can be observed that CBP-TDF has a narrower range of prediction errors compared with CBP and ARIMA. As shown in Fig. 4.7, when the process is under operating condition II, the CBP is able to follow the trend when ARR has sharp and rapid fluctuations. Obvious discrepancies occur when ARR rises or drops sharply. However, the CBP-TDF method can still perform well under this situation. The

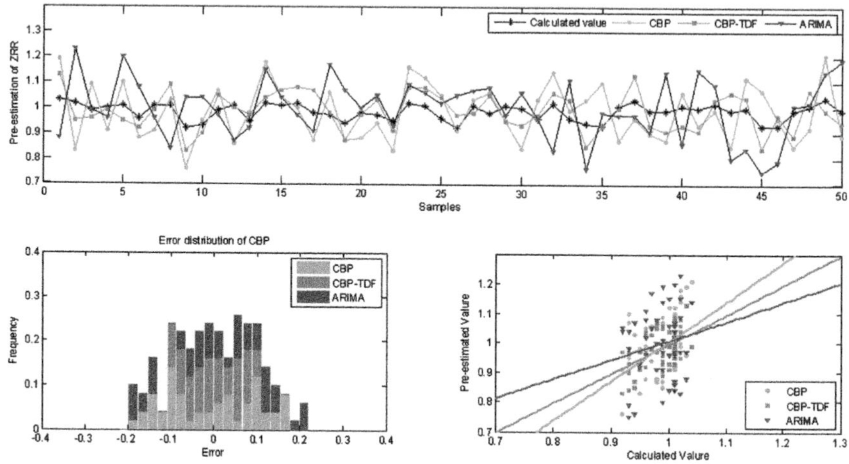

FIGURE 4.8 ARR estimation results when using the MPC approach under operating condition I.

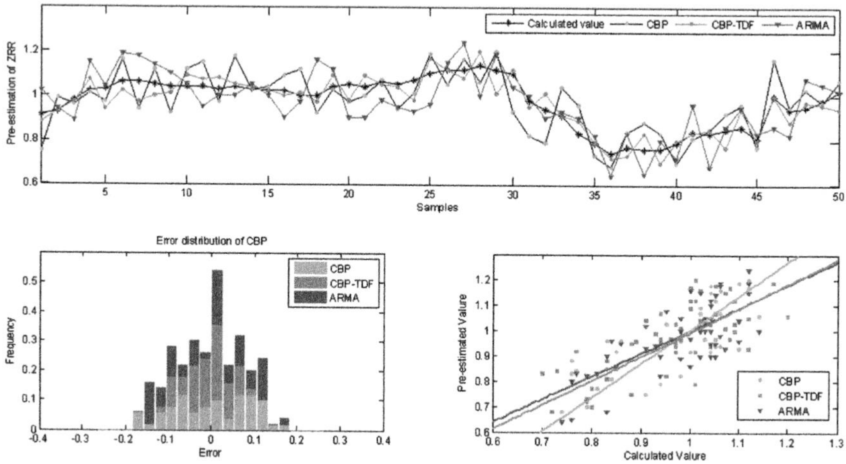

FIGURE 4.9 ARR estimation results when using the MPC approach under operating condition II.

distributions of the prediction error and the similarities between the calculations and predictions also prove that CBP-TDF performs well, compared with the other prediction algorithms. Figs. 4.8 and 4.9 illustrate the estimation results of ARR in the MPC approach. Similarly to the AMC case, it can be observed that the ARR estimated by the CBP-TDF algorithm is also closer to the actual value.

Tables 4.5 and 4.6 show the statistical indices of the prediction performance under all four situations. For the AMC approach, when the process is under operating condition II, the proposed algorithm is mostly close to the actual value. When the process is under operating condition II, MAE and RMSE of the proposed algorithm increased from 0.024 and 0.032 to 0.039 and 0.049, respec-

TABLE 4.5 Comparison of ARR estimation results for the AMC approach.

Operating condition	Prediction method	MAE	DR	RMSE	p-value
I	CBP	0.071	[−0.16, 0.14]	0.083	0.5606
	CBP-TDF	0.024	[−0.07, 0.07]	0.032	0.8449
	ARIMA	0.097	[−0.20, 0.18]	0.112	0.7819
II	CBP	0.089	[−0.24, 0.25]	0.098	0.7847
	CBP-TDF	0.039	[−0.15, 0.13]	0.049	0.9861
	ARIMA	0.113	[−0.23, 0.22]	0.131	0.4760

TABLE 4.6 Comparison of ARR estimation results for the MPC approach.

Operating condition	Prediction method	MAE	DR	RMSE	p-value
I	CBP	0.071	[−0.19, 0.17]	0.106	0.3377
	CBP-TDF	0.024	[−0.10, 0.10]	0.061	0.9701
	ARIMA	0.097	[−0.19, 0.17]	0.117	0.9605
II	CBP	0.089	[−0.17, 0.17]	0.094	0.7659
	CBP-TDF	0.039	[−0.11, 0.08]	0.049	0.8906
	ARIMA	0.113	[−0.15, 0.16]	0.117	0.8313

tively. Under both operating conditions, the performance of CBP-TDF is better than that of the other algorithms. The advantage of CBP-TDF is also observed in the MPC approach. The MAE and RMSE of CBP-TDF are the least under the two operating conditions. In addition, the deviation ranges of the proposed algorithm are narrower than the others. However, the performance of CBP-TDF in the MPC approach is a little poorer than in the AMC approach. An explanation is that the changes of the coefficients which are used in the MPC and not considered in CBP-TDF have influence on the prediction result. A t-test was performed to estimate the differences between the predicted and actual ARR. The p-values indicate that there are no significant differences between the predicted and actual ARR for all prediction algorithms. In addition, the predicted values obtained by CBP-TDF have less differences with the actual ARR.

In CBP-TDF, the values of the coefficients and weights in the similarity calculation are listed in Table 4.7, where α_1, α_2, α_3, and α_4 are the weights for ORP, temperature, flow rate, and additive dosage in the trend similarity calculation, respectively; w_1, w_2, w_3, w_4, w_5, and w_6 are the weights for ORP, temperature, flow rate, additive dosage, and the inlet and outlet copper ion concentrations in process variable similarity calculation, respectively. It is observed that the weights of ORP (α_1 and w_1) under operating condition I are higher than those under operating condition II. This is mainly because the ORP could reveal the underlying change of copper removal reactions. In addition, under operat-

TABLE 4.7 Optimized weights and coefficients in ARR prediction of two control approaches.

Control approach	Operating condition	$[\alpha_1, \alpha_2, \alpha_3, \alpha_4]$	$[w_1, w_2, w_3, w_4, w_5, w_6]$	λ
AMC approach	I	[0.25, 0.10, 0.21, 0.24]	[0.25, 0.09, 0.14, 0.20, 0.19, 0.13]	0.6
	II	[0.27, 0.07, 0.22, 0.24]	[0.28, 0.06, 0.16, 0.20, 0.20, 0.10]	0.47
MPC approach	I	[0.24, 0.10, 0.23, 0.23]	[0.23, 0.07, 0.16, 0.21, 0.20, 0.13]	0.59
	II	[0.27, 0.06, 0.23, 0.25]	[0.28, 0.07, 0.19, 0.20, 0.19, 0.07]	0.43

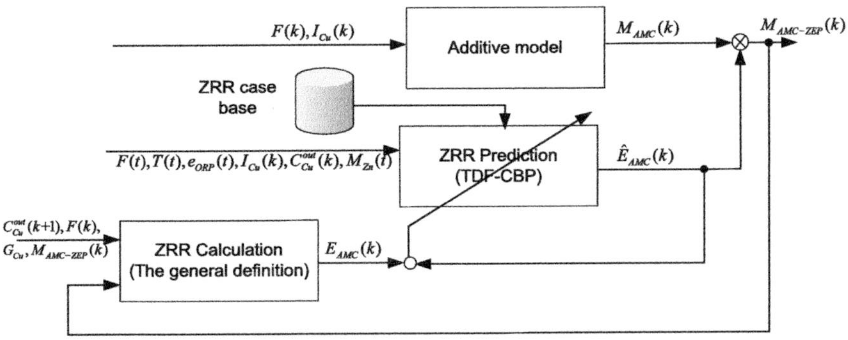

FIGURE 4.10 The additive model setting integrated with zinc requirement ratio prediction.

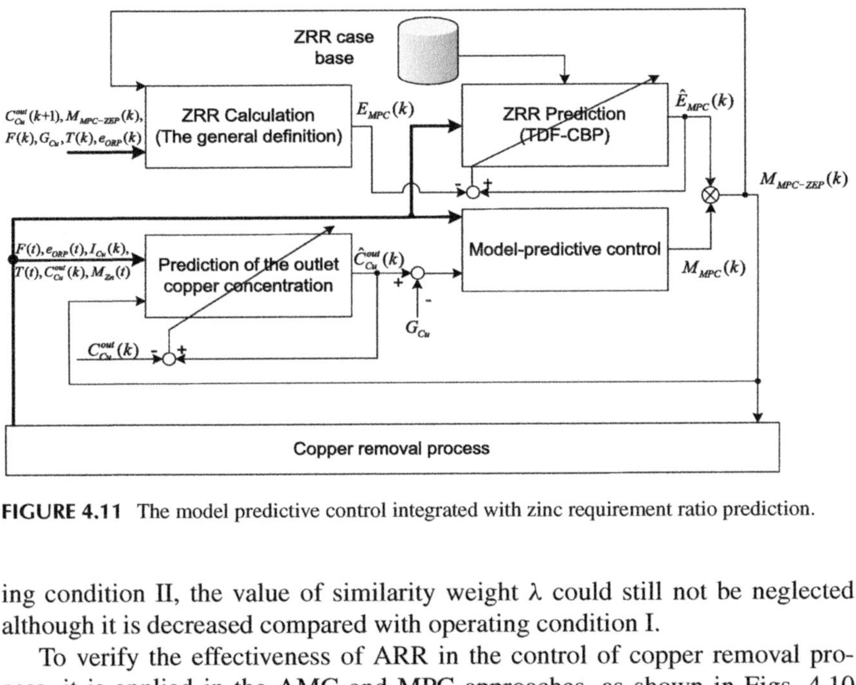

FIGURE 4.11 The model predictive control integrated with zinc requirement ratio prediction.

ing condition II, the value of similarity weight λ could still not be neglected although it is decreased compared with operating condition I.

To verify the effectiveness of ARR in the control of copper removal process, it is applied in the AMC and MPC approaches, as shown in Figs. 4.10 and 4.11. ARR is predicted online using CBP-TDF and used to correct the additive dosage given by the AMC or MPC approach. The actual ARR is calculated using the actual zinc powder consumption, the outlet copper ion concentration, and related process variables. The difference between the predicted ARR and the calculated ARR is used to adjust the ARR prediction model. To evaluate the control performance, three indices are introduced, which include qualification ratio, fluctuation range, and median. The qualification ratio is calculated as

$$R_Q = \frac{N_Q}{N_A} \times 100\%, \tag{4.24}$$

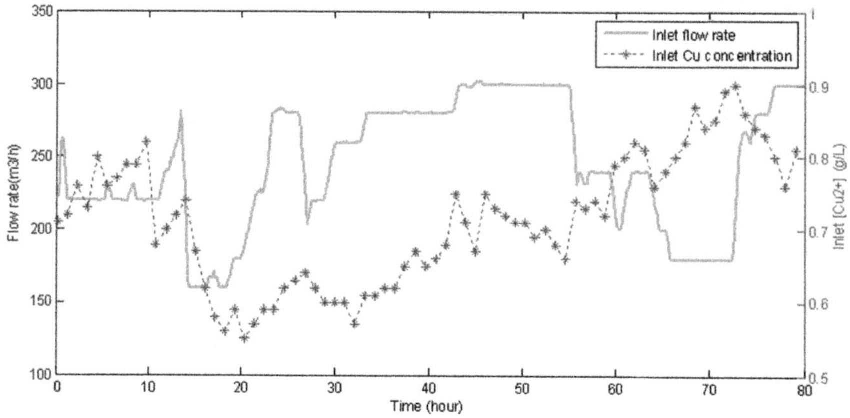

FIGURE 4.12 Inlet copper ion concentration and flow rate.

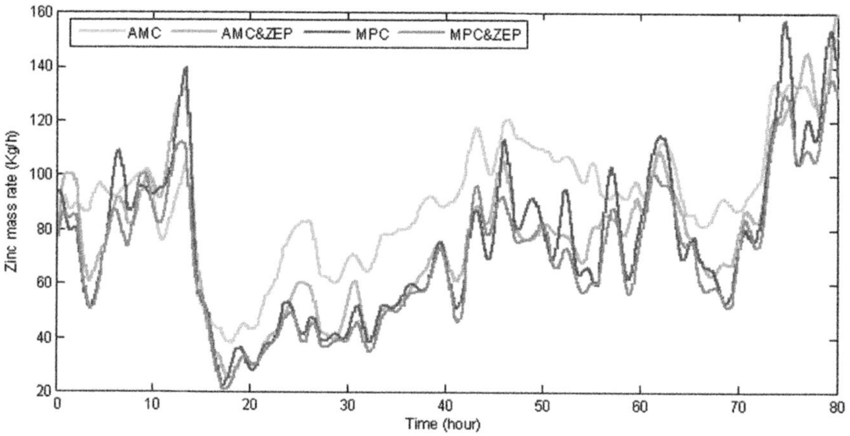

FIGURE 4.13 Comparison of zinc powder dosage.

where N_Q denotes the number of samples whose outlet copper ion concentration is located in the range specified by the production requirement and N_A denotes the number of all the samples.

The simulation results of the four control approaches, including AMC, AMC integrated with zinc requirement ratio prediction (AMC-ZRRP), MPC, and MPC integrated with zinc requirement ratio prediction (MPC-ZRRP), are shown in Figs. 4.12–4.15. Fig. 4.12 shows the copper ion concentration and flow rate of the inlet solution during the simulation. Fig. 4.13 shows the additive dosages given by the four control approaches. Figs. 4.14 and 4.15 present the value and statistical analysis of the resulting outlet copper ion concentration using the four control approaches. It is observed that the outlet copper ion concentrations of the AMC approach are mostly lower than the lower limit of the outlet copper ion

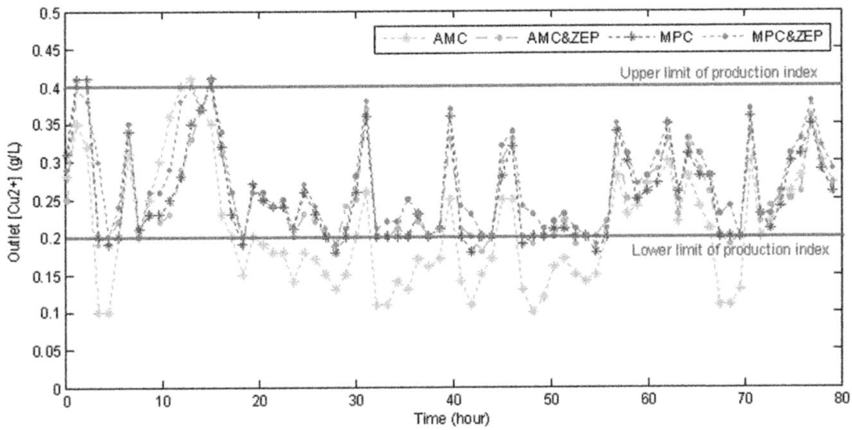

FIGURE 4.14 Comparison of the outlet copper ion concentration.

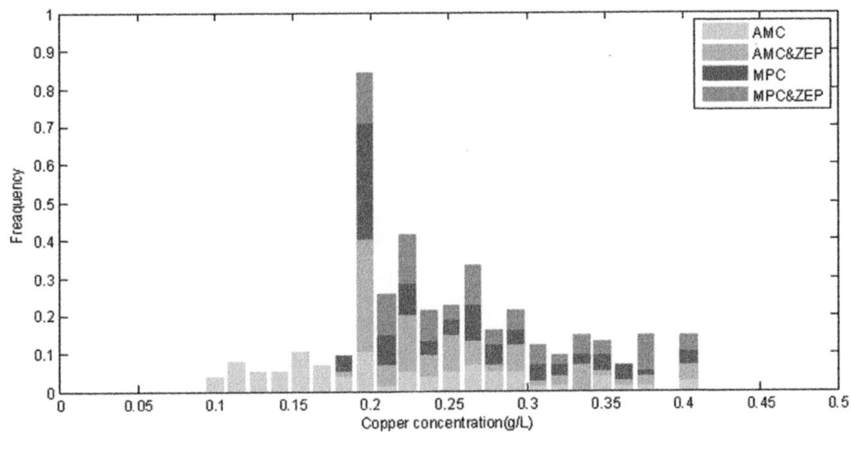

FIGURE 4.15 Frequencies of the outlet copper ion concentration under four control approaches.

concentration. However, in the AMC-ZRRP approach, the amounts of zinc powder are less than those in the AMC approach most of the time, which keeps the outlet copper ion concentration in the specified range. The zinc powder dosage given by MPC is close to that given by AMC-ZRRP. However, the MPC approach changes more sharply and greatly than AMC-ZRRP, which is easier to make the process unstable. The MPC-ZRRP approach can adjust the dosage of zinc powder smoothly and timely, and nearly all the outlet copper ion concentrations are qualified. To sum up, the qualified ratios of the outlet copper ion concentrations of AMC and MPC are increased by 27% and 6%, respectively (Table 4.8). Moreover, the adjustment of zinc powder dosage can adapt to the underlying change of inlet and reaction conditions, which proves the effectiveness of the proposed approach.

TABLE 4.8 Comparison of ARR estimation results for the MPC approach.

Approach	QR (%)	FR (g/L)	M (g/L)	AZM (kg/L)	p-value
AM	63	[0.10, 0.41]	0.20	90	8.71×10^{-4}
AM-ZRRP	90	[0.18, 0.40]	0.25	77	
MPC	88	[0.18, 0.41]	0.25	74	0.0948
MPC-ZRRP	96	[0.19, 0.40]	0.27	70	

References

[1] I.M. Ahmed, Y.A. El-Nadi, J.A. Daoud, Cementation of copper from spent copper-pickle sulfate solution by zinc ash, Hydrometallurgy 110 (1–4) (2011) 62–66.

[2] N. Demirkıran, A. Ekmekyapar, A. Künkül, A. Baysar, A kinetic study of copper cementation with zinc in aqueous solutions, International Journal of Mineral Processing 82 (2) (2007) 80–85.

[3] F.I. Gamero, J. Meléndez, J. Colomer, Process diagnosis based on qualitative trend similarities using a sequence matching algorithm, Journal of Process Control 24 (9) (2014) 1412–1424.

[4] B. Zhang, C. Yang, Y. Li, X. Wang, H. Zhu, W. Gui, Additive requirement ratio prediction using trend distribution features for hydrometallurgical purification processes, Control Engineering Practice 46 (2016) 10–25.

[5] K. Laatikainen, M. Lahtinen, M. Laatikainen, E. Paatero, Copper removal by chelating adsorption in solution purification of hydrometallurgical zinc production, Hydrometallurgy 104 (1) (2010) 14–19.

[6] S. Pellegrini, E. Ruiz, A. Espasa, Prediction intervals in conditionally heteroscedastic time series with stochastic components, International Journal of Forecasting 27 (2) (2011) 308–319.

[7] K. Villez, C. Rosén, F. Anctil, C. Duchesne, P.A. Vanrolleghem, Qualitative representation of trends (QRT): extended method for identification of consecutive inflection points, Computers & Chemical Engineering 48 (2013) 187–199.

[8] R. Rengaswamy, T. Hägglund, V. Venkatasubramanian, A qualitative shape analysis formalism for monitoring control loop performance, Engineering Applications of Artificial Intelligence 14 (1) (2001) 23–33.

[9] B. Sun, W. Gui, T. Wu, Y. Wang, C. Yang, An integrated prediction model of cobalt ion concentration based on oxidation–reduction potential, Hydrometallurgy 140 (2013) 102–110.

[10] J.C. Wong, K.A. Mcdonald, A. Palazoglu, Classification of process trends based on fuzzified symbolic representation and hidden Markov models, Journal of Process Control 8 (1998) 395–408.

[11] M. Oliveira, R.M. Crujeiras, A. Rodriguezcasal, A plug-in rule for bandwidth selection in circular density estimation, Computational Statistics & Data Analysis 56 (12) (2012) 3898–3908.

[12] J. Vanlaer, G. Gins, J.F.M. Van Impe, Quality assessment of a variance estimator for Partial Least Squares prediction of batch-end quality, Computers & Chemical Engineering 52 (2013) 230–239.

[13] B. Zhang, C. Yang, H. Zhu, Y. Li, W. Gui, Kinetic modeling and parameter estimation for competing reactions in copper removal process from zinc sulfate solution, Industrial & Engineering Chemistry Research 52 (48) (2013) 17074–17086.

[14] Y. Xie, S. Xie, X. Chen, W. Gui, C. Yang, L. Caccetta, An integrated predictive model with an on-line updating strategy for iron precipitation in zinc hydrometallurgy, Hydrometallurgy 151 (2015) 62–72.

[15] Z. Liang, Y. Li, S. Xia, Adaptive weighted learning for linear regression problems via Kullback-Leibler divergence, Pattern Recognition 46 (4) (2013) 1209–1219.

Chapter 5

Real-time adjustment of zinc powder dosage based on fuzzy logic

Contents

5.1 Copper removal performance evaluation based on ORP

In order to precisely control the copper ion concentration to its set-point, the zinc powder dosage has to be determined accurately to account for the actual need of copper removal [1][2]. In practice, the operators evaluate the operating condition and adjust the zinc powder dosage according to the inlet and outlet copper ion concentrations. However, with the discrete-delayed measurement of copper ion concentrations, the operation is not able to follow the variation of inlet and outlet copper ion concentrations in time, which may lead to instability of the process [3]. According to the kinetics of multielectrode reactions, ORP, which is related to the reaction rate, is a real-time external indicator of the internal reaction state. Therefore, ORP can be utilized in the monitoring and adjusting of the copper removal process [4][5][6].

5.1.1 Relationship between copper ion concentration and ORP

Before evaluating the copper removal performance based on ORP, the relationship between ORP and the outlet copper ion concentration has to be analyzed. The copper removal process mainly involves two oxidation–reduction reactions:

Modeling, Optimization, and Control of Zinc Hydrometallurgical Purification Process
https://doi.org/10.1016/B978-0-12-819592-5.00015-6

(i) Copper cementation: $CuSO_4 + Zn \rightarrow ZnSO_4 + Cu$.
(ii) Copper comproportionation: $CuSO_4 + Cu + H_2O \rightarrow Cu_2O \downarrow + H_2SO_4$.

The two reactions follow first-order chemical reaction dynamics, whose reaction rate can be expressed as

$$r_i = -k_i C, \tag{5.1}$$

where r_i is the reaction rate of the ith main reaction ($i = 1, 2$), C is the outlet copper ion concentration, and k_i is the rate constant of the ith main reaction ($i = 1, 2$). Based on the independence principle of coexisting chemical reactions and mass balance principle, the reaction rate of copper ions can be expressed as

$$\frac{dC}{dt} = r_1 + r_2 = -(k_1 + k_2)C. \tag{5.2}$$

According to the Arrhenius equation, the activation energy $E_{a \cdot i}$ can affect the rate constant k_i:

$$k_i = A_i \exp\left(-\frac{E_{a \cdot i}}{RT}\right), \tag{5.3}$$

where A_i and $E_{a \cdot i}$ are the preexponential factor and activation energy of the ith main reaction, respectively, R is the gas constant, and T is the absolute temperature.

From the perspective of the electrode reaction, the main electrochemical reaction on the anode is

$$Zn \rightarrow Zn^{2+} + 2e^-, \qquad E^0 = -0.736 \text{ V}. \tag{5.4}$$

On the cathode, the reactions include

$$Cu^{2+} + 2e^- \rightarrow Cu, \qquad E^0 = 0.337 \text{ V}, \tag{5.5}$$

$$2Cu^{2+} + H_2O + 2e^- \rightarrow Cu_2O + H^+, \qquad E^0 = 0.203 \text{ V}. \tag{5.6}$$

According to the independency principle of parallel electrode reactions, these reactions share a common electrode potential. The electrode potential, also known as mixed potential, determines the reaction rate by affecting the activation energy or the electron transfer rate between oxidant and reductant [7][8].

According to the kinetics of the electrode reaction, the actual activation energies of the anode and cathode reactions can be formulated as [9]

$$\begin{cases} E_{ea} = E_{ea}^0 - (1-\lambda)nF(e_{mix} - e_{eq}), \\ E_{ec} = E_{ec}^0 - \lambda nF(e_{mix} - e_{eq}), \end{cases} \tag{5.7}$$

where E_{ea}^0 and E_{ec}^0 are the standard activation energy of the anode and cathode electrode reactions, respectively, e_{mix} is the mixed potential, e_{eq} is the standard equilibrium potential, n is the number of electron moles, F is the Faraday constant, and λ and $(1-\lambda)$ are the coefficients describing the effects of electrode potential variation on the activation energy of the cathode and anode electrode reactions [9].

There are two opinions on the relationship between mixed potential and ORP. One is that ORP and mixed potential exhibit a linear relationship:

$$e_{mix} = pe_{orp} - q, \tag{5.8}$$

where p and q are parameters to be identified. The identification result indicates that p is close to 1 and q is close to 0. The other opinion is that ORP equals the mixed potential. Then, by combining Eq. (5.2) and Eq. (5.8), the mathematical relationship between ORP and copper ion concentration is

$$\ln\left(\frac{C_t}{C_0}\right) = t\left[A_1\exp\left(-\frac{E_{e\cdot 1} + 2\lambda F(pe_{orp} + q - e_{eq\cdot 1})}{RT}\right) + A_2\exp\left(-\frac{E_{e\cdot 2} + \lambda F(pe_{orp} + q - e_{eq\cdot 2})}{RT}\right)\right], \tag{5.9}$$

where R, F, $e_{eq\cdot 1}$, and $e_{eq\cdot 2}$ are constant; the values of C_0 and T are obtained by measurement; $E_{e\cdot 1}$, $E_{e\cdot 2}$, λ, A_1, A_2, p, and q are unknown and need to be identified by minimizing the difference between the predicted outlet copper ion concentration ($C_{kinetic}$) and the measured concentration (C_{real}):

$$\min \frac{1}{n_T}\sum_{k=1}^{n_T}[C_{kinetic}(k) - C_{real}(k)]^2, \tag{5.10}$$

where n_T is the number of identification samples.

5.1.2 ORP-based process evaluation

Analyzing ORP is a feasible approach to evaluate the reaction state of the copper removal process [6]. However, copper removal is a complex process with various influencing factors. Other impurities in the solution can also affect the reaction state, which introduces uncertainties in the relationship between ORP and copper removal performance. To handle these uncertainties, fuzzy logic is selected for process evaluation [10][11][12].

ORP is an indicator of the current state of the copper removal process, while the variation of ORP indicates the trend of the copper removal performance [13]. If we only utilize the current value of ORP in process evaluation, then the obtained evaluation result is inaccurate, which misguides the process operation. For example, if the ORP is more positive than its normal upper limit, then the process is considered to be in a serious state (extreme shortage of zinc powder) and more zinc powder is required. However, if the inlet copper ion concentration decreases sharply and the required zinc powder dosage is decreased accordingly, then the process will enter another serious state (excessive zinc powder). Therefore, both ORP and its variation trend should be utilized in the process evaluation.

In the evaluation process, ORP and its trend are classified into several fuzzy sets with different fuzzy membership functions. The kinetic model and the production limitation of the concentrations are used to determine the parameters of these membership functions. After the fuzzification of ORP and its trend, a set of fuzzy inference rules is adopted to evaluate the process performance [14].

The evaluation schedule mainly consists of four stages. The first stage is preparing the input variables. A linear regression method is applied to extract the real-time trend of ORP. The trend extraction includes the following steps.

***Step 1*:** Determine the initial window size of the time series as W, and set the threshold θ.

***Step 2*:** In the time window, build a linear model to approximate the time series of ORP:

$$\hat{x}(k) = a(k - k_0) + b, \tag{5.11}$$

where k is the time step in the window, k_0 is the initial time step, $\hat{x}(k)$ is the approximated value, and a and b are parameters of the linear regression model, whose values are determined by minimizing the sum of least-squares:

$$S = \sum_{i=0}^{W} (\hat{x}(k_i) - x(k_i))^2, \tag{5.12}$$

where $x(k_i)$ is the actual value of ORP at time step k_i in the original window. The fitted parameter a represents the initial ORP trend. The singular value decomposition method is used to account for the matrix singularity.

***Step 3*:** Add a new data point to the window.

***Step 4*:** Calculate the sum of least-squares with the fitted model over the new windows. If S is less than θ, then repeat step 3. Otherwise, the new data point will be treated as the initial data of the new window, and step 2 is repeated.

In stage 2, the input variables are converted to linguistic variables, which is known as fuzzification. Three types of membership functions are used in this stage.

***Type* 1:** The generalized bell membership function:

$$\mu_i^j(\chi) = \frac{1}{1 + |\left(x - \delta_i^j\right)/w_i^j|^{2m}},$$

where δ_i^j and w_i^j ($i = -3, \cdots, 3,\ j = 1, 2$) are the center and amplitude of the curve, respectively, and m is a positive constant.

***Type* 2:** Z-shaped membership function:

$$\varphi^j(\chi) = \begin{cases} 1, & x \le \alpha^j, \\ 1 - 2\left(\frac{\chi-\alpha^j}{\beta^j-\alpha^j}\right)^2, & \alpha^j \le \chi \le \frac{\alpha^j+\beta^j}{2}, \\ 2\left(\frac{\chi-\beta^j}{\beta^j-\alpha^j}\right)^2, & \frac{\alpha^j+\beta^j}{2} \le \chi \le \beta^j, \\ 0, & \chi \ge \beta^j, \end{cases}$$

where both α^j and β^j are parameters deciding the slope of the curve $\varphi^j(\chi)$.

***Type* 3:** S-shaped membership function:

$$\psi^j(\chi) = \begin{cases} 0, & x \le c^j, \\ 2\left(\frac{\chi-c^j}{d^j-c^j}\right)^2, & c^j \le \chi \le \frac{c^j+d^j}{2}, \\ 1 - 2\left(\frac{\chi-d^j}{d^j-c^j}\right)^2, & \frac{c^j+d^j}{2} \le \chi \le d^j, \\ 1, & \chi \ge d^j, \end{cases}$$

where both c^j and d^j are parameters deciding the slope of the curve $\psi^j(\chi)$.

According to the reaction kinetics, the increase of ORP indicates an increasing copper ion concentration and vice versa. However, the reflection is vague. The fuzzified value should be of physical meaning and suitable for control. Therefore, the definition of those membership functions should take the limitation of the outlet copper ion concentration into account. Denote the limitation of the outlet copper ion concentration as $[L_{MIN}, L_{MAX}]$. The parameters in the ORP membership function are calculated according to the limitations by using $e_{orp} = g(C_{Cu^{2+}})$, which is the inverse function of Eq. (5.9). To make the evaluation more intuitive, the fuzzy sets are labeled by fuzzy languages indicating reaction states of the copper removal process. If the outlet copper ion concentration is at the center of production limitations and ORP is in a stable situation, then it is abbreviated as S. Similarly, if ORP is relatively low or very low, it is labeled as little low (LL) or very low (VL). If ORP is relatively high or very high, it is labeled as little high (LH) or very high (VH).

The trend of ORP indicates its underlying variation, which determines the future value of ORP. A sharp trend can force ORP to return to a stable situation from an unstable one within minutes. It can also cause the deviation of ORP from an originally stable situation within seconds. Therefore, the parameters of the ORP trend membership functions are determined according to the influence of the ORP trend on the ORP state. Denote the time threshold for ORP trend analysis as τ(min). If ORP skips or drops from one condition to a higher or lower situation during τ minutes, its trend is viewed as low positive or low negative. If ORP skips or drops from a stable situation to the highest or lowest situation, its trend is labeled as high positive or high negative. The relevant

TABLE 5.1 Parameter calculation and fuzzy language definition for ORP.

Label	Meaning	Function	Calculation of parameters
VL	Very low	$\varphi^1(x)$	$\alpha^1 = g(L_{MIN}),\ \beta^1 = \alpha^1 - p_2\omega_{-1}^1$
LL	Little low	$\mu_{-3}^1(x)$	$\sigma_{-1}^1 = g(q_1 L_{MIN} + (1-q_1)L_{MAX}),\ \omega_{-1}^1 = p_1(\sigma_0^1 - \alpha^1),\ m=2$
S	Stable	$\mu_0^1(x)$	$\sigma_0^1 = g(q_0 L_{MIN} + (1-q_0)L_{MAX}),\ \omega_0^1 = p_0(\sigma_1^1 - \sigma_{-1}^1),\ m=2$
LH	Little high	$\mu_1^1(x)$	$\sigma_1^1 = g(q_1 L_{MIN} + (1-q_1)L_{MAX}),\ \omega_1^1 = p_1(d^1 - \sigma_0^1),\ m=2$
VH	Very high	$\psi^1(x)$	$d^1 = g(L_{MAX}),\ c^1 = d^1 + p_2\omega_1^1$

TABLE 5.2 Parameter calculation and fuzzy language definition for the trend of ORP.

Grade	Meaning	Function	Calculation of parameters
HN	High negative	$\varphi^2(x)$	$\alpha^2 = (\alpha^1 - d^1)/\tau,\ \beta^2 = \alpha^2 - r_2\omega_{-1}^2$
LN	Low negative	$\mu_{-1}^2(x)$	$\sigma_{-1}^2 = (\sigma_{-1}^1 - \sigma_1^1)/\tau,\ \omega_{-1}^2 = r_1(\sigma_0^2 - \alpha^2),\ m=2$
Z	Zero	$\mu_0^1(x)$	$\sigma_0^2 = 0,\ \omega_0^2 = r_0(\sigma_1^2 - \sigma_{-1}^2),\ m=2$
LP	Low positive	$\mu_1^1(x)$	$\sigma_1^2 = (\sigma_1^1 - \sigma_{-1}^1)/\tau,\ \omega_1^2 = r_1(d^2 - \sigma_0^2),\ m=2$
HP	High positive	$\psi^2(x)$	$d^2 = (d^1 - \alpha^1)/\tau,\ c^2 = d^2 + r_2\omega_1^2$

information of fuzzification of ORP and its trend are presented in Tables 5.1 and 5.2. Fig. 5.1(a and b) shows the shapes of the memberships.

In stage 3, a set of rules is applied on the fuzzy sets obtained in stage 2 for evaluating the copper removal process. Both ORP and its trend are considered in the process evaluation. If both ORP and its trend are in an acceptable situation without fluctuation, then the copper removal process is considered in a stable state. If ORP is in an acceptable situation, however, with a sharp trend, then the process is not as stable as the value of ORP indicated. Similarly, if ORP exceeds its normal limitation, however, with a good trend leading to an acceptable situation, then the process situation is not as bad as it seems to be [15]. Therefore, according to this idea, the fuzzy inference rules for process evaluation are formulated, as shown in Fig. 5.1(c) [16]. The evaluation linguistic variables are obtained by the Max-Prod operator following the Larsen fuzzy inference method [17].

In stage 4, the linguistic variables are converted into a single numerical value, i.e., the evaluation grade. The centroid defuzzification method is applied where the crisp value of the output variable is computed by finding the center of gravity of the membership function for the fuzzy value [18][19]. The output value is calculated as

$$y^* = \frac{\sum_{i=1}^{n} \mu(y_i) y_i}{\sum_{i=1}^{n} \mu(y_i)}. \tag{5.13}$$

A more positive grade means a situation with a higher concentration and vice versa. An evaluation grade closer to zero indicates a more stable process con-

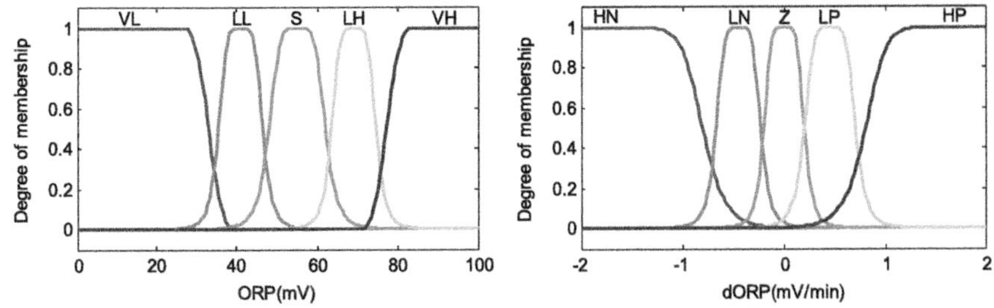

(a) Membership of ORP

(b) Membership of ORP trend

ORP / dORP	VL	LL	S	LH	VH
HP	ML	LL	LH	MH	VH
LP	ML	LL	S	MH	VH
Z	VL	LL	S	LH	VH
LN	VL	ML	S	LH	MH
HN	VL	ML	LL	LH	MH

(c) the fuzzy interference rules

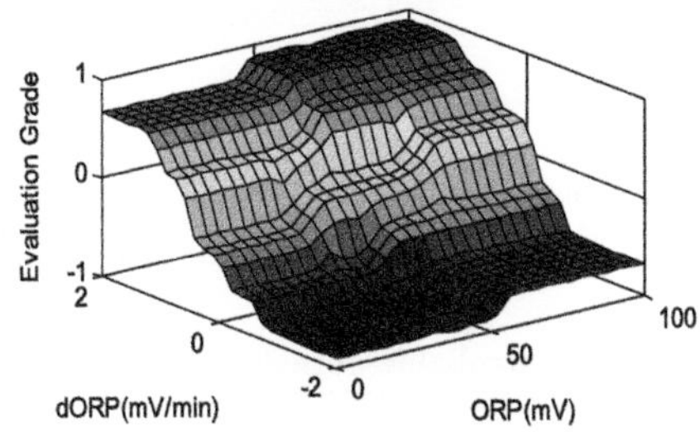

(d) the relationships among the evaluation grade, ORP, and its trend

FIGURE 5.1 Fuzzy logic process evaluation based on ORP.

dition. Fig. 5.1(d) shows the evaluation grade surface of the process, which displays the relationship between two inputs (ORP and its trend) and the response output (evaluation grade).

5.2 Controllable domain-based fuzzy rule extraction for copper removal

In this section, a fuzzy rule extraction approach based on the concept of controllable domain (CD-FRE) is introduced. The fuzzy rule extraction approach mainly includes two parts, as shown in Fig. 5.2. First, the controllable domain, which indicates the achievable control effect under an operating condition, is classified according to the evaluation grade of the copper removal process. Then, the fuzzy control rules are extracted for each domain based on a fuzzified SVM [20].

5.2.1 Data preparation

The chemical reactors in the copper removal process have large volumes, which give rise to the cumulative effect. There exists a certain time delay between control and response [21]. Therefore, before rule extraction, the production data are preprocessed. Denoting the reactor volume and flow rate as v and x_1, respectively, the time delay can be computed approximately as $T_D = \alpha v / x_1$, where α

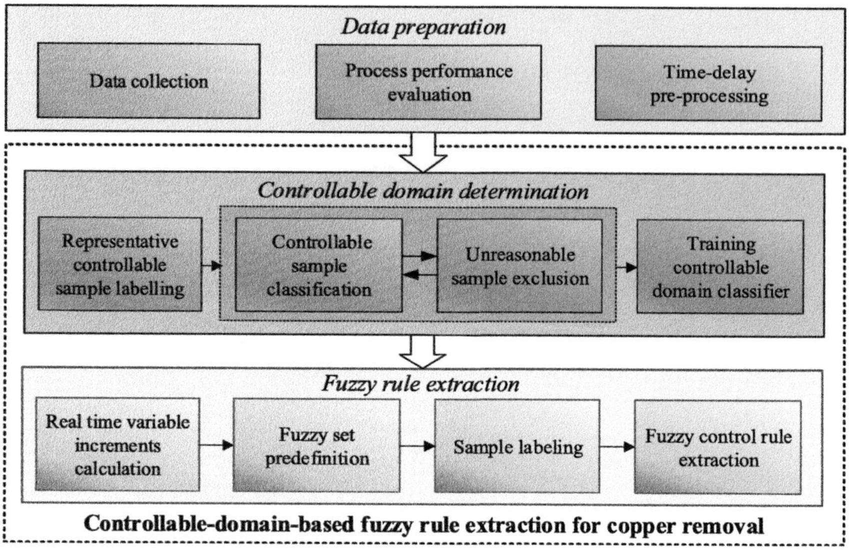

FIGURE 5.2 The framework for the controllable domain-based fuzzy rule extraction.

TABLE 5.3 Input variables for fuzzy rule extraction.

Variable name	Detection method	Symbol	Input variables
Flow rate [m^3/h]	Online	x_1	$d^0, d^{-1}, \cdots, d^{-\Gamma}$
ORP [mV]	Online	x_2	$d^0, d^{-1}, \cdots, d^{-\Gamma}$
Zinc powder dosage (#1) [kg/h]	Online	x_3	$d^0, d^{-1}, \cdots, d^{-\Gamma}$
Zinc powder dosage (#2) [kg/h]	Online	x_4	$d^0, d^{-1}, \cdots, d^{-\Gamma}$
Inlet Cu^{2+} concentration [g/L]	Offline	x_5	d^{-1}
Outlet Cu^{2+} concentration [g/L]	Offline	x_6	d^{-1}

is a coefficient. The times of lag taps could be calculated as $\Gamma = T_D/T_S$, where T_S is the sampling interval of the real-time process variables.

Therefore, the selection of input variables for rule extraction should take the time delay into consideration, as shown in Table 5.3. The selected input variables include discrete-delayed measurements of inlet and outlet copper concentrations and the current and lagged values of online measurable process variables. The lagged value of a process variable can be represented by the combination of the process variable and a backward operator d^{-i}, i.e., $x_j(t-i) = x_j(t)d^{-i}$.

Besides the preprocessing of time delay, ORP and its trend are transformed to the evaluation grade using the method introduced in Section 5.1 using a fuzzy logic technique: $g = f_{\text{EVAL}}(x_2, dx_2)$, where x_2 and dx_2 denote ORP and its first derivative, respectively, and g and $f_{\text{EVAL}}(\bullet)$ are the evaluation grade and a fuzzy function converting ORP into g, respectively.

5.2.2 Controllable domain determination

Due to the diversity of operators' experience, they give different operations under similar operating conditions, resulting in distinct copper removal performance. Therefore, the operation records should be screened and classified before fuzzy rule extraction, which includes three steps: representative controllable sample labeling, controllable sample classification and exclusion, and controllable domain classification.

1) Representative controllable sample labeling: Denote the set of input variables as $\mathbf{I} = \{\mathbf{I}_R, \mathbf{I}_M\}$, $\mathbf{I}_R \in R^{n\times 4(\Gamma+1)}$, $\mathbf{I}_M \in R^{n\times 4}$, where $\mathbf{I}_R = \{\mathbf{x}_1 d^0, \cdots, \mathbf{x}_1 d^\Gamma, \cdots, \mathbf{x}_4 d^0, \cdots, \mathbf{x}_4 d^\Gamma\}$ and $\mathbf{I}_M = \{\mathbf{x}_5, \mathbf{x}_5 d^{-1}, \mathbf{x}_6, \mathbf{x}_6 d^{-1}\}$. First, the data samples in data set $\mathbf{I}$ are labeled. Copper removal is a continuous production process. The variation of the operation performance during an interval of certain length can be used to evaluate the control operation. Several types of control performance are defined by the value and variation of the evaluation grade during the observation interval, as shown in Fig. 5.3.

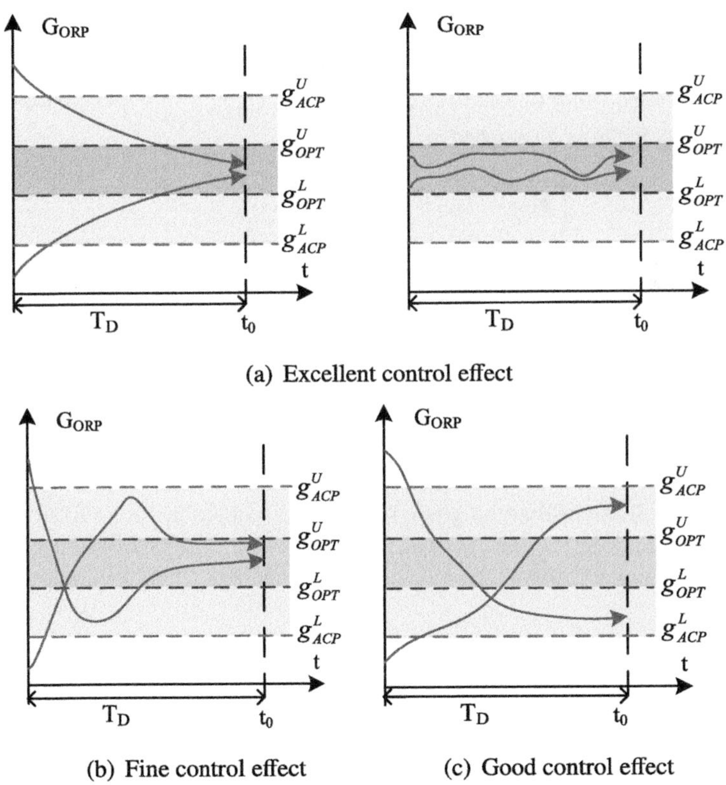

FIGURE 5.3 Typical movements of evaluation grades for a data sample with different controllable labels.

Definition 1. If the evaluation grade keeps stable in the optimal range from $t_0 - T_D$ to t_0, which can be described as

$$\begin{cases} g(t) \in [g_O^L, g_O^U], & t \in [t_0 - t_D, t_0], \\ |g'(t)| < \dfrac{g_O^L + g_O^U}{T_D}, & t \in [t_0 - t_D, t_0], \end{cases}$$

where g_O^L and g_O^U are the lower and upper limits of the optimal range of the evaluation grade, respectively, or when $g(t)$ reaches the optimal range from an nonoptimal value with no fluctuations, which can be described as

$$\begin{cases} g(t_0 - T_D) \notin [g_O^L, g_O^U], \\ g(t_0) \in [g_O^L, g_O^U], \\ g'(t)[x_2(t_0 - T_D) - 0.5(g_O^L + g_O^U)] < 0, t \in [t_0 - T_D, t_0], \end{cases}$$

then the corresponding data sample is labeled as an excellent control effect sample.

Definition 2. If the evaluation grade reaches the optimal range from a nonoptimal range during the interval $[t_0 - T_D, t_0]$ and stays in the acceptable range after T_S, which can be described as

$$\begin{cases} g(t_0 - T_D) \notin [g_O^L, g_O^U], \\ g(t) \in [g_A^L, g_A^U], t \in [t_0 - T_S, t_0], \\ g(t_0) \in [g_O^L, g_O^U], \end{cases}$$

where g_A^L and g_A^U are the lower and upper limits of the acceptable range of evaluation grade, respectively, and T_S is the time the evaluation grade reaches the acceptable range for the first time, then the corresponding data sample is labeled as a fine control effect sample.

Definition 3. If the evaluation grade reaches the acceptable range from the unacceptable range during the interval $[t_0 - T_D, t_0]$ and stays in the acceptable range after being acceptable, which can be described as

$$\begin{cases} g(t_0 - T_D) \notin [g_O^L, g_O^U], \\ |g'(t)| < \dfrac{g_A^L + g_A^U}{T_D}, t \in [t_0 - T_S, t_0], \\ g(t_0) \in [g_A^L, g_A^U] \cap \overline{[g_O^L, g_O^U]}, \end{cases}$$

then the corresponding data sample is labeled as a good control effect sample.

2) Controllable sample classification and exclusion: Due to the complex operating conditions and diversity of operators' experience, only a part of the samples can be identified using the proposed definitions. In addition, the irrational operation samples must be excluded. Therefore, the sample selection process is labor-intensive. An improved positive and unlabeled (PU) learning approach is applied to classify samples and exclude poor samples [22][23][24][25].

The kNN-PU algorithm has beens designed for two-class classification problems [26]. It cannot be directly applied to classify the operation data samples into multiple controllable domains. Therefore, the multiclass data sample classification problem is transferred to a series of two-class problems. First, some samples with excellent control effect are labeled. The kNN-PU algorithm is used to identify all the samples with excellent control effect from the data set. These samples form the excellent controllable domain. Before further classification of the remaining samples, the samples, whose input characters are similar to excellent samples but have a different control effect, are excluded. Then, all the samples with fine control effect are selected to construct the fine controllable domain. Similarly, the samples with good control effect are selected to construct the good controllable domain. The remaining unclassified samples are used to construct the marginal controllable domain. Fig. 5.4 and Algorithm 1 show the procedure of controllable domain classification.

In Algorithm 1, the process condition similarity (PCS) is expressed as

$$\mathrm{Sim}_{PC}(\mathbf{u}_i,\mathbf{y}_j) = \frac{\sum_{k=1}^{m} w_k^2 \times \mathbf{u}_{i,k} \times \mathbf{y}_{j,k} - \alpha}{\sqrt{\sum_{k=1}^{m}(w_k \mathbf{u}_{i,k})^2 - \beta^x}\sqrt{\sum_{k=1}^{m}(w_k \mathbf{y}_{i.k})^2 - \beta^y}}, \quad (5.14)$$

where

$$\alpha = \sum_{a=0}^{\Gamma} w_{m1,a}^2 \times (m_{1,i}^x d^{-a} \times m_{1,j}^x d^{-a} + m_{2,i}^x d^{-a} \times m_{2,j}^x d^{-a}), \quad (5.15)$$

$$\beta^x = \sum_{a=0}^{\Gamma}\left[w_{m1,a}^2 (m_{1,i}^x d^{-a})^2 + (m_{2,i}^x d^{-a})^2\right], \quad (5.16)$$

$$\beta^y = \sum_{a=0}^{\Gamma}\left[w_{m1,a}^2 (m_{1,i}^y d^{-a})^2 + (m_{2,i}^y d^{-a})^2\right], \quad (5.17)$$

where $\mathbf{u}_i$ is the ith sample of excellent/fine/good control effect sample set $(\mathbf{s}_i^1, \mathbf{s}_i^2, \mathbf{s}_i^3)$, y_j is the jth sample of the first/second/third left sample set $(\mathbf{s}_i^{'}, \mathbf{s}_i^{''}, \mathbf{s}_i^{'''})$, $\mathbf{u}_{i,k}$ and $\mathbf{y}_{j,k}$ are the kth variables of $\mathbf{u}_i$ and $\mathbf{y}_j$, respectively, w_k is the weight of the kth variable, $m_{1,i}^x$ and $m_{2,i}^x$ are the zinc amounts of the first and second reactors of $\mathbf{u}_i$, respectively, $m_{1,i}^y$ and $m_{2,i}^y$ are the zinc amounts of the first and second reactors of $\mathbf{y}_j$, respectively, and $w_{m1,a}$ ($w_{m1,a} \in w = \{w_1, ..., w_n\}$) and $w_{m2,a}$ ($w_{m2,a} \in w = \{w_1, ..., w_n\}$) are the weights of $m_1 d^{-a}$ and $m_2 d^{-a}$, respectively.

Algorithm 1 PU learning based on kNN and SVM.

Input: $\mathbf{I} = \{\mathbf{I}_R, \mathbf{I}_m\} = \{s_1, \cdots, s_{m0}\}^T$, the industrial sample set, θ_{SIM}, threshold

Output: The controllable domains

```
Label the "excellent" controllable samples from I according to Definition 1
Apply the kNN-PU learning algorithm to classify all the "excellent" controllable samples from I, and denote the "excellent" controllable and the left sample sets as C1 = {s_1^1, ..., s_{m1}^1}^T and S1 = {s_1^{L1}, ..., s_{m2}^{L1}}^T (m1 + m2 = m0), respectively
for each "excellent" controllable sample s_i^1 do
    for each left sample s_j^{L1} do
        Calculate the PCS value Sim_PC(s_i^1, s_j^{L1})s_j^{L1}
        if Sim_PC(s_i^1, s_j^{L1}) > θ_SIM then
            reject s_j^{L1} from S1
Obtain a new left sample set S1* = {s_1^{LN1}, ..., s_{m3}^{LN1}}^T (m3 ≤ m2)
Label the "fine" controllable samples from S1*, according to Definition 2
Apply the kNN-PU learning algorithm to classify all the "fine" controllable samples from S1*, and denote the "fine" controllable and the left sample sets as C2 = {s_1^2, ..., s_{m4}^2}^T and S2 = {s_1^{L2}, ..., s_{m5}^{L2}}^T, respectively (m4 + m5 = m3)
for each "fine" controllable sample s_i^2 do
    for each left sample s_j^{L2} do
        Calculate the PCS value Sim_PC(s_i^2, s_j^{L2})
        if Sim_PC(s_i^2, s_j^{L2}) > θ_SIM then
            reject s_j^{L2} from S2
Obtain a new left sample set S2* = {s_1^{LN2}, ..., s_{m6}^{LN2}}^T (m6 ≤ m5)
Label the "good" controllable samples from S2*, according to Definition 3
Apply the kNN-PU learning algorithm to classify all the "good" controllable samples from S2*, and denote the "good" controllable and the left sample sets as C3 = {s_1^3, ..., s_{m7}^3}^T and S3 = {s_1^{L3}, ..., s_{m8}^{L3}}^T (m7 + m8 = m6), respectively
for each "good" controllable sample S_i^3 do
    for each left sample s_j^{L3} do
        Calculate the PCS value Sim_PC(s_i^3, s_j^{L3})
        if Sim_PC(s_i^3, s_j^{L3}) > θ_SIM then
            reject s_j^{L3} from S3
Obtain a new third left industrial set, and take this set as the "marginal" controllable sample set C4 = {s_1^4, ..., s_{m9}^4}^T (m9 ≤ m8)
```

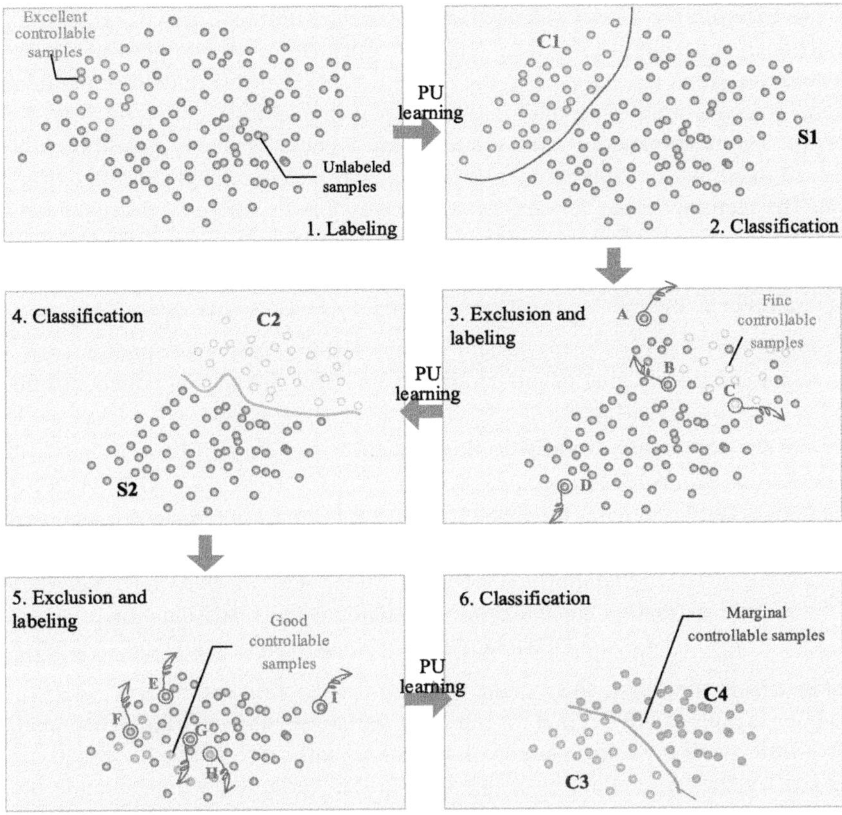

FIGURE 5.4 The procedure of controllable domain classification based on kNN-PU learning (A to I are the rejected irrational-operated samples, S1 and S2 are the first and second left sample sets, respectively, and C1, C2, C3, and C4 are the excellent, fine, good, and marginal controllable sample sets, respectively).

3) Controllable domain classification: By using Algorithm 1, the operation data samples are tagged and divided into different categories. However, this classification approach is not appropriate for online identification of new samples. The samples in the controllable domains differ from the samples on which these classifiers are trained. Therefore, a new controllable domain classifier should be trained using multiclass SVM with the controllable samples.

5.2.3 Rule extraction using a fuzzified SVM classifier

Generally, the control rules are different under different controllable domains. Therefore, rule extraction should be conducted for each domain. The fuzzy rule extraction includes four major steps: (1) variable increment calculation; (2) fuzzy set predefinition; (3) sample labeling; and (4) fuzzy rule extraction based on classification.

(1) Variable increment calculation: In practice, the zinc powder dosage is determined according to the flow rate and copper ion concentration of the inlet solution. However, only using the discrete-delayed measurement of the inlet copper ion concentration cannot realize the real-time adjustment of zinc powder dosage. The increments of online measurable process variables should then be utilized to adjust the zinc powder dosage, which are calculated for rule extraction. The increments are denoted as $\Delta x_i = x_i - x_i d^{-1}$, where x_i $(i = 1, \cdots, 4)$ represents online measurable process variables. In the step, the historical values of these variables are replaced by their increments. As a result, the ith data sample in the jth controllable domain $(j = 1, \cdots, 4)$, $s_i^j = \{x_{i,1}^j, \cdots, x_{i,1}^j d^{\Gamma}, \cdots, x_{i,4}^j, \cdot, x_{i,4}^j d^{\Gamma}, x_{i,5}^j, x_{i,5}^j d^{-1}, x_{i,6}^j, x_{i,6}^j d^{-1}\}$, is transformed to a sample with the present value of input variables in Table 5.3 and increments of the online measurable variables, i.e., $s_j^c = \{x_{1,j}, \cdots, \cdots_{6,j}, \Delta x_{1,j}, \cdots, \Delta x_{4,j}\}$. If s_j^c is in the excellent controllable domain, then $c = E,\ j = 1, \cdots, m_E$. If s_j^c is in the fine controllable domain, then $c = F,\ j = 1, \cdots, m_F$. If s_j^c is in the good controllable domain, then $c = G,\ j = 1, \cdots, m_G$. If s_j^c is in the marginal controllable domain, then $c = M,\ j = 1, \cdots, m_M$.

(2) Fuzzy set predefinition: Generally, fuzzy sets have to be predefined in fuzzy rule extraction. In the proposed method, the Gaussian functions are adopted to fuzzify the input variables, e.g., $\mu(x) = \exp\left(-\frac{x-o}{2\delta^2}\right)$, where o is the distance from the origin and δ is the width of the function curve.

Four fuzzy sets are defined for the input variables, which include very small (VS), little small (LS), little large (LL), and very large (VL). Five fuzzy sets are defined for the increments of online measurable variables, which include obvious decrease (OD), slight decrease (SD), roughly stable (RS), slight increase (SI), and obvious increase (OI). For the jth sample, the membership degrees of x_i $(i = 1, \cdots, 6)$ for VS, LS, LL, and VL and Δx_i $(i = 1, \cdots, 4)$ for OD, SD, RS, SI, and OI are denoted as $\mu_{\text{VS},i}(j)$, $\mu_{\text{LS},i}(j)$, $\mu_{\text{LL},i}(j)$, $\mu_{\text{VL},i}(j)$, and $\mu_{\text{OD},i}(j)$, $\mu_{\text{SD},i}(j)$, $\mu_{\text{RS},i}(j)$, $\mu_{\text{SI},i}(j)$, and $\mu_{\text{OI},i}(j)$. The parameters of these membership functions can be obtained based on the fuzzy Shannon entropy:

$$\begin{aligned}
\max & -\sum_{j=1}^{N_i}\left\{\mu_{\text{FS},i}(j)\log\mu_{\text{FS},i}(j) + (1-\mu_{\text{FS},i}(j))\log(1-\mu_{\text{FS},i}(j))\right\}, \\
\text{s.t.} \quad & \mu_{\text{FS},i}(j) \in [0, 1] \quad \forall x_{i,j}, \\
& o_{\text{FS},i} \in [o_{\text{FS},i}^L, o_{\text{FS},i}^U], \\
& \sigma_{\text{FS},i} \in [\sigma_{\text{FS},i}^L, \sigma_{\text{FS},i}^U],
\end{aligned} \tag{5.18}$$

where FS denotes the type of fuzzy set that could be VS, LS, LL, VL, OD, SD, RS, SI, and OI, $o_{\text{FS},i}$ and $\delta_{\text{FS},i}$ are parameters of membership function $\mu_{\text{FS},i}(j)$, $o_{\text{FS},i}^L$ and $o_{\text{FS},i}^U$ are the lower and upper bounds of the parameter $o_{\text{FS},i}$, and $o_{\text{FS},i}^L$ and $o_{\text{FS},i}^U$ are the lower and upper bounds of parameter $\delta_{\text{FS},i}$. In practice, $o_{\text{FS},i}^L$

and $o_{\mathrm{FS},i}^{U}$ are set in accordance with the distribution of the training data samples:

$$\begin{cases} o_{\mathrm{VS}}^{i} \in [min\{x_i\}, Q_{\alpha 1}(x_i)], \\ o_{\mathrm{LS}}^{i} \in [Q_{\alpha 1}(x_i), Q_{\alpha 2}(x_i)], \\ o_{\mathrm{LL}}^{i} \in [Q_{\alpha 2}(x_i), Q_{\alpha 3}(x_i)], \\ o_{\mathrm{VL}}^{i} \in [Q_{\alpha 3}(x_i), max\{x_i\}], \end{cases} \tag{5.19}$$

where o_{VS}^{i}, o_{LS}^{i}, o_{LL}^{i}, and o_{VL}^{i} are the centers of the VS, LS, LL, and VL membership functions, $Q_{\alpha i}$ $(i = 1, 2, 3)$ is the quartile of the distribution of x_i, and $Q_{\alpha 1}$, $Q_{\alpha 2}$, and $Q_{\alpha 3}$ are the first, second, and third quartiles, respectively. The parameter constraints of the membership function for Δx_i are determined as

$$\begin{cases} o_{\mathrm{OD}}^{i} \in [\min(x_i) - Q_{3/4}(x_i)/\Delta T_D, Q_{1/4}(x_i) - Q_{3/4}(x_i)/\Delta T_D], \\ o_{\mathrm{SD}}^{i} \in [Q_{1/4}(x_i) - Q_{1/2}(x_i)/\Delta T_D, Q_{1/4}(x_i) - Q_{3/4}(x_i)/\Delta T_D], \\ o_{\mathrm{RS}}^{i} \in [-Q_{1/4}(x_i)/\Delta T_D, Q_{1/4}(x_i)/\Delta T_D], \\ o_{\mathrm{SI}}^{i} \in [Q_{1/2}(x_i) - Q_{1/4}(x_i)/\Delta T_D, Q_{3/4}(x_i) - Q_{1/4}(x_i)/\Delta T_D], \\ o_{\mathrm{OI}}^{i} \in [Q_{3/4}(x_i) - Q_{1/4}(x_i)/\Delta T_D, \min(x_i) - Q_{1/4}(x_i)/\Delta T_D]. \end{cases} \tag{5.20}$$

The range of $\delta_{\mathrm{FS},i}$ is determined by the limits of $o_{\mathrm{FS},I}$:

$$\delta_{\mathrm{FS}}^{i} \in \left[0.05(o_{\mathrm{FS},i}^{U} - o_{\mathrm{FS},i}^{L}), 0.3(o_{\mathrm{FS},i}^{U} - o_{\mathrm{FS},i}^{L})\right]. \tag{5.21}$$

(3) Sample labeling: In this step, the data samples s_j^c are labeled in accordance with the memberships of the zinc dosage increments in the next control step ($\Delta x_3 p^1$ and $\Delta x_4 p^1$, where p^1 is the forward shift operator, $\Delta x_i p^1 = \Delta x_i(t+1) - \Delta x_i(t)$). Firstly, the fuzzy sets that $\Delta x_3 p^1$ and $\Delta x_4 p^1$ belong to are determined by obtaining their maximum membership degree:

$$C_{\Delta x_{i,j}} = \begin{cases} \mathrm{OD}, & f(\mu) = \mu_{\mathrm{OD},i}(j), \\ \mathrm{SD}, & f(\mu) = \mu_{\mathrm{ID},i}(j), \\ \mathrm{RS}, & f(\mu) = \mu_{\mathrm{RS},i}(j), \\ \mathrm{SI}, & f(\mu) = \mu_{\mathrm{SI},i}(j), \\ \mathrm{OI}, & f(\mu) = \mu_{\mathrm{OI},i}(j), \end{cases} \tag{5.22}$$

where $f(\mu) = \max\{\mu_{\mathrm{OD},i}(j), \mu_{\mathrm{SD},i}(j), \mu_{\mathrm{RS},i}(j), \mu_{\mathrm{SI},i}(j), \mu_{\mathrm{OI},i}(j)\}$. Then, the data samples are labeled by evaluating the overall variation of zinc powder dosage. The larger the decrease of zinc powder dosage quantity per unit time, the more negative the sample label is, and vice versa. After screening of the unreasonable and duplicate combinations, the sample labels are reduced to 11 types, as shown in Fig. 5.5.

Δx_4 \ Δx_3	LD	SD	O	SI	LI
LD	-5	-4	-3	—	—
SD	-4	-2	-1	—	—
O	-3	-1	0	1	3
SI	—	—	1	2	4
LI	—	—	3	4	5

FIGURE 5.5 Classes determined by the combination of adjustment amounts of zinc powder added in the first and second reactors ("-" indicates an unreasonable combination).

(4) Fuzzy rule extraction: By using the rules illustrated in Fig. 5.5, the data samples of the four controllable domains can be labeled with zinc powder dosage. Then the multiclass SVM can be used in each controllable domain to obtain the support vectors for fuzzy rule extraction. The fuzzy rule extraction includes the following steps.

a) Label the sample $s_j^c = \{x_{1,j}, \cdots, x_{6,j}, \Delta x_{1,j}, \cdots, \Delta x_{4,j}\}$ with $y = i$, $i = -5, \cdots, 5$.

b) Train the multiclass SVM, f_{SVM}^c, for each controllable domain $\mathbf{S}^c = s_1^c, \cdots, s_{m_c}^c$.

c) Select the support vectors $\mathbf{SV}_i^c$ $(i = 1, \cdots, m_{SV}^c)$ from the controllable domains and project $s_{SV,i}^c$ in the coordinate axes to obtain $\mathbf{s}_{SV,i}^c$.

d) Divide $\{x_{1,j}, \cdots, x_{6,j}\}$ and $\{\Delta x_{1,j}, \cdots, \Delta x_{4,j}\}$ into fuzzy and five fuzzy sets, respectively, by using the defined membership functions.

e) Choose the fuzzy set $C_{i,j\max}^c$ providing maximum membership degree, $\max_j \left\{\mu_{i,j}^{c,\mathrm{SV}}\right\}$, for each attribute of $\mathbf{s}_{\mathrm{SV},i}^c$.

f) Generate fuzzy rules by support vectors as follows:

i) Denote $C_{i,j\max}^c \in \{\mathrm{VS, LS, LL, VL}\}$ and $C_{\Delta i,j\max}^c \in \{\mathrm{OD, SD, RS, SI, OI}\}$ as the fuzzy sets with the highest membership degree for x_i $(i = 1, \cdots, 6)$ and $\Delta x_i (i = 1, \cdots, 4)$ of $\mathbf{s}_{\mathrm{SV},i}^c$.

ii) Let $C_{sv,j}$ be the class of the corresponding support vector $\mathbf{SV}_i^c$ classified by multiclass SVM classes.

iii) Represent the rule R generated by $\mathbf{SV}_i^c$ (or the projected sample $\mathbf{s}_{\mathrm{SV},i}^c$) as follows:
If x_1 is $C_{x_1,jmax}^c$, ..., x_6 is $C_{x_6,jmax}^c$ and Δx_1 is $C_{x_1,jmax}^c$, ..., Δx_4 is $C_{\Delta x_4,jmax}^c$, then $\mathbf{s}_{\mathrm{SV},i}^c$ belongs to $C_{\mathbf{sv},j}$.

iv) When the label of $\mathbf{sv}_j^c$ is changed to the corresponding fuzzy classes of $\Delta x_3 p^1$ and $\Delta x_3 p^1$, the formulation rule is translated into:
If x_1 is $C_{x_1,jmax}^c$, ..., x_6 is $C_{x_6,jmax}^c$ and Δx_1 is $C_{x_1,jmax}^c$, ..., Δx_4 is $C_{\Delta x_4,jmax}^c$, then $\Delta x_3 p^1$ is $C_{\Delta x_3 p^1,j}$ and $\Delta x_4 p^1$ is $C_{\Delta x_4 p^1,j}$.

5.3 Results

A total of 23,106 samples collected from a plant were used to verify the effectiveness and performance of the proposed approach. These samples were classified into four controllable domains. The numbers of samples and value ranges of each controllable domain are listed in Table 5.4. The distributions of variables in the sample are obtained, as shown in Figs. 5.6 and 5.7.

TABLE 5.4 Numbers and bounds of controllable domains for the copper removal process.

Domains	Excellent	Fine	Good	Marginal
N_D*	5012	6452	6843	4012
x_1	[100, 310]	[100, 310]	[100, 310]	[100, 310]
x_2	[39, 86]	[39, 95]	[19, 110]	[19, 130]
dx_2	[−1.25, 2]	[−3.4, 2]	[−5, 4.8]	[−11.2, 2]
x_3	[31, 162]	[31, 181]	[31, 181]	[13, 29]
x_4	[13, 28]	[13, 27]	[13, 28]	[3.9, 4.2]
x_5	[0.55, 0.94]	[0.55, 0.94]	[0.55, 0.94]	[0.55, 0.94]
x_6	[0.1, 0.4]	[0.1, 0.6]	[0.1, 0.6]	[0.1, 0.6]

* N_D is the number of the samples in the controllable domains.

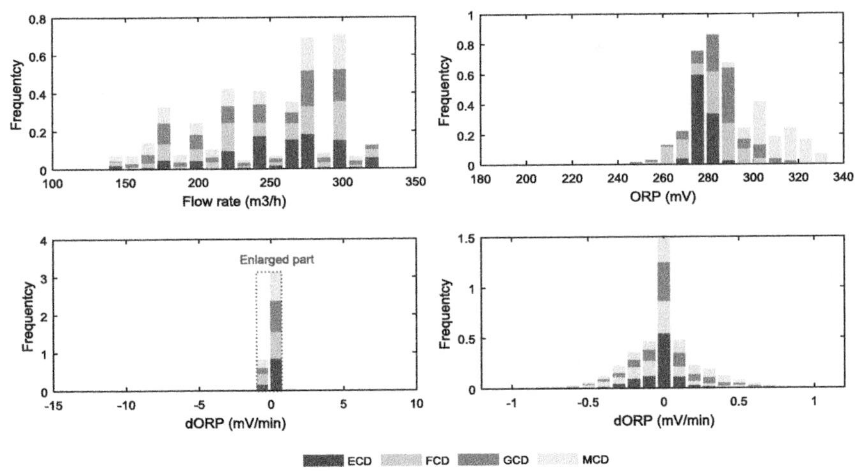

FIGURE 5.6 Distributions of inlet flow rate, ORP, ORP increments, and enlarged part of ORP in the range [−0.8, 0.8] for four controllable domains.

As can be observed from Table 5.4, the fluctuation ranges of the flow rate and inlet copper ion concentration are the same for all controllable domains. However, the ranges of the remaining variables increase with change of controllable domains. More specifically, the increase direction is from excellent, fine, good,

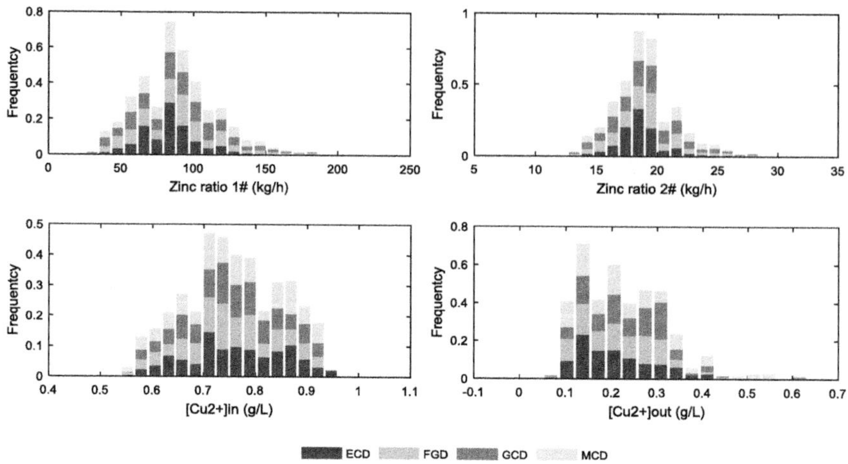

FIGURE 5.7 Distributions of zinc amounts added in the first reactor and the second reactor, the distributions of the inlet and outlet copper concentrations for four controllable domains.

to marginal controllable domains. As can be observed from Figs. 5.6 and 5.7, the variations of the distribution center of most variables across all the controllable domains are small. However, the distributions of ORP in different controllable domains are distinguishable from each other. The distribution centers of ORP in excellent, fine, good, and marginal controllable domains are 60, 78, 78, and 89 mV, respectively. Therefore, if ORP is in an acceptable range, then it is easy to obtain an expected control effect. Fig. 5.7 indicates that the distribution center of the outlet copper ion concentration (0.15 g/L) is much less than the optimal concentration (0.3 g/L) in the excellent controllable domain. This phenomenon indicates that there is a preference for human operators to add excessive zinc powder.

After domain classification, a set of fuzzy rules is extracted for each domain. To verify the effectiveness of the CD-FRE strategy, manual operation (MO) and two common rule extraction methods, including a common fuzzy rule extraction method based on SVM (CFRE) and a PU learning-based fuzzy rule extraction method (PU-FRE), are selected for comparison purposes. The purpose is to prove the usefulness of the improved CD-FRE strategy. For the CFRE method, the rules are extracted for the whole sample set without classification and exclusion. For the PU-FRE method, the data samples are classified into four domains without sample exclusion, and four sets of rules are extracted for each controllable domain. The criteria for rule extraction include accuracy and coverage:

1) Accuracy: The accuracy of rule R associated with the class C_s, $C_s \in \{5, 4, 3, 2, 1, 0, 1, 2, 3, 4, 5\}$, is defined as $A^R_{\mathrm{CS}} = \sum_{i=1}^{mcs} \mu_R(X_i) / \sum_{j=1}^{m} \mu_R(X_j)$, where m_{CS} is the number of data samples belonging to C_s, m is the total number of data samples, and μ_R is the product of the membership degrees

TABLE 5.5 Results of fuzzy rule extraction for the copper removal process.

Rule extraction method		Numbers of rules	Accuracy (%)	Coverage (%)
Excellent	PU-FRE	10	79	96
	CD-FRE	10	79	96
Fine	PU-FRE	12	73	90
	CD-FRE	13	77	92
Good	PU-FRE	14	71	84
	CD-FRE	13	73	89
Marginal	PU-FRE	9	70	74
	CD-FRE	8	71	79
CFRE		14	73	78

of data sample to each fuzzy set that belongs to the antecedent of rule R; μ_R is formulated as $\mu_R(X_i) = \prod_{j=1}^{n_{\text{input}}} \mu C_{x_{j,i}}(x_{j,i})$, where $n_{rminput}$ is the number of input variables, $x_{j,i}$ is the jth variable of X_i, and $\mu C_{x_{j,i}}$ is the membership degree of $x_{j,i}$; $x_{j,i}$ belongs to $C_{x_{j,i}}$, which is the fuzzy class that appears in rule R.

2) Coverage: Fuzzy coverage is a measurement for the ability of rule R to affect the patterns, which is defined as $CO_{CS}^{R} = \sum_{j=1}^{m} \mu_R(X_i)/m$.

Table 5.5 lists the rule prediction results of different rule extraction methods. It can be observed that the fuzzy rules extracted perform well in excellent and fine controllable domains. However, in the other two controllable domains, the prediction accuracy and coverage decrease. This is due to the increased complexity and diversity of reaction conditions in the last two controllable domains. Therefore, the accuracy of the rules is influenced. The performance of fuzzy rules extracted using PU-FRE and CD-FRE is similar in the excellent controllable domain. However, the accuracy and coverage of the rules extracted using PU-FRE decrease in other controllable domains. Therefore, for marginal controllable domain, the performance of rules extracted by CFRE is similar to that of the rules extracted by CD-FRE.

The adjustment actions are shown in Figs. 5.8 and 5.9. Fig. 5.8 presents the values of the flow rate, ORP, and inlet copper ion concentration. These data are sampled from four representative inlet conditions. In the first 10 hours, the flow rate changed several times, while the inlet copper ion concentration was kept stable. During 10–20 hours, the flow rate was kept stable, while the inlet copper ion concentration was fluctuating. During 30–40 hours, both flow rate and inlet copper ion concentration were kept stable. After 30 hours, both flow rate and inlet copper ion concentration changed frequently. Fig. 5.8 also presents the adjustment of the zinc powder dosage. It can be observed that the adjustment frequency of MO is lower than that of the other approaches. The adjustment amplitude changes largely in some cases. The adjustment actions given by CFRE and PU-FRE are frequent. However, by using the two approaches, the

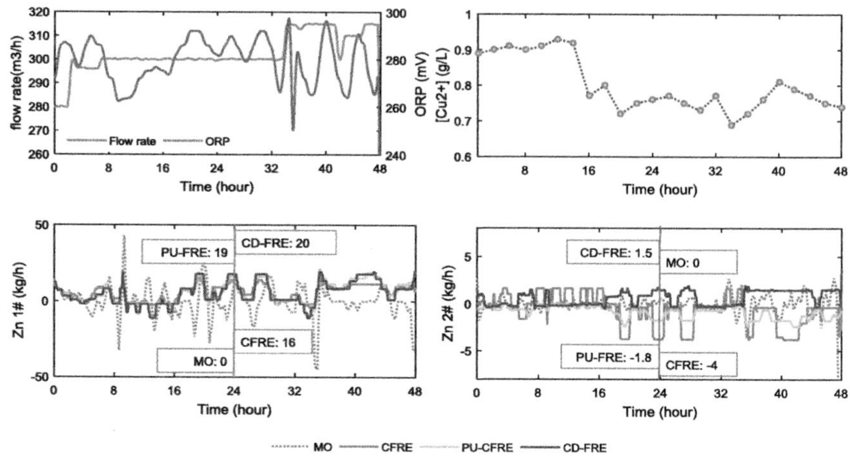

FIGURE 5.8 The inlet flow rate and ORP, the inlet copper ion concentration, and the adjustments of the zinc powder dosage for the first and second reactors.

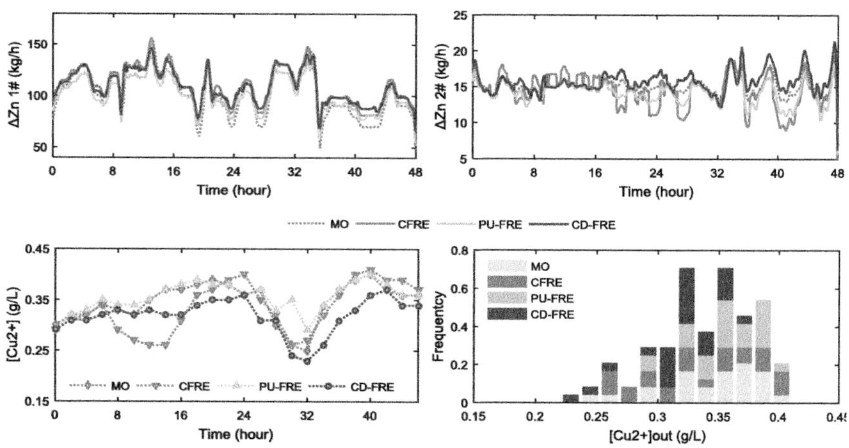

FIGURE 5.9 The zinc powder dosage in the first and second reactors and the value curves and distributions of the outlet copper ion concentration under four different control approaches.

adjustments for the first and second reactors are unreasonable and conflicting in some cases. For example, at the 24th hour, the zinc powder dosage of reactor #1 is increased, however, it is decreased for reactor #2. These conflicting operations, which are common in the original operation data set, cannot be rejected by CFRE and PU-FRE. The data produced by the unreasonable operations reduce the expected control efficiency of the extracted rules. Therefore, the operations derived using the proposed approach are timely and proper.

Fig. 5.9 presents the zinc powder dosage. The zinc powder dosage of reactor #1 adjusted by CFRE, PU-FRE, and CD-FRE changed more smoothly than

TABLE 5.6 Evaluation indices for the control results.

Methods	Oscillation range (g/L)	Q_r (g/L)	Mean value (g/L)	Median (g/L)
MO	[0.25, 0.41]	95%	0.35	0.36
CFRE	[0.26, 0.41]	91%	0.33	0.33
PU-FRE	[0.28, 0.405]	95%	0.36	0.38
CD-FRE	[0.23, 0.37]	100%	0.31	0.32

Q_r is qualification rate.

when adjusted by MO. However, the adjustment actions of CD-FRE for reactor #2 are more moderate and reasonable compared with CFRE, PU-FRE, and MO, which is more beneficial for the stability of the process.

The value and distribution of the outlet copper ion concentration under the three approaches are also shown in Fig. 5.9. Under the three approaches, the variation trajectories of the outlet copper ion concentration are similar in trend. However, the outlet copper ion concentration trajectory driven by CD-FRE is closer to the center of the required range (0.3 g/L), which indicates the stability of process. The distribution of the outlet copper ion concentration using CD-FRE is also the narrowest. On the other hand, as shown in Table 5.6, four indices are introduced to quantify the control performance, including oscillation range, qualification rate, mean value, and median. For the MO approach, the median and mean values of the outlet copper ion concentration are higher than the optimal value 0.3 g/L in most cases. In addition, the outlet copper ion concentration is not always qualified. For the CFRE approach, the qualification ratio of the outlet copper ion concentration is slightly decreased. However, the median and mean values of the outlet copper ion concentration are closer to 0.3 g/L. For the PU-FRE approach, the fluctuation range of the outlet copper ion concentration is narrowed. However, the median and mean values of the outlet copper ion concentration are away from 0.3 g/L. For the CD-FRE approach, it is superior to MO, CFRE, and PU-FRE on all the four indices. It could not only remove the impurity effectively, but also reserve an appropriate amount of copper ions as activator for the cobalt removal stage.

References

[1] B. Zhang, C. Yang, H. Zhu, Y. Li, W. Gui, Kinetic modeling and parameter estimation for competing reactions in copper removal process from zinc sulfate solution, Industrial & Engineering Chemistry Research 52 (48) (2013) 17074–17086.

[2] K. Laatikainen, M. Lahtinen, M. Laatikainen, E. Paatero, Copper removal by chelating adsorption in solution purification of hydrometallurgical zinc production, Hydrometallurgy 104 (1) (2010) 14–19.

[3] M. Ruano, J. Ribes, A. Seco, J. Ferrer, Low cost-sensors as a real alternative to on-line nitrogen analysers in continuous systems, Water Science and Technology 60 (12) (2009) 3261–3268.

[4] R. Yu, F. Chi, W. Cheng, J. Chang, Application of pH, ORP, and DO monitoring to evaluate chromium(VI) removal from wastewater by the nanoscale zero-valent iron (nZVI) process, Chemical Engineering Journal 255 (2014) 568–576.
[5] C. Antileo, H. Medina, C. Bornhardt, C. Muñoz, F. Jaramillo, J. Proal, Actuators monitoring system for real-time control of nitrification–denitrification via nitrite on long term operation, Chemical Engineering Journal 223 (2013) 467–478.
[6] S. Lackner, H. Horn, Evaluating operation strategies and process stability of a single stage nitritation–anammox SBR by use of the oxidation–reduction potential (ORP), Bioresource Technology 107 (2012) 70–77.
[7] C. Chang, Y. Ma, C. Lo, Application of oxidation–reduction potential as a controlling parameter in waste activated sludge hydrolysis, Chemical Engineering Journal 90 (3) (2002) 273–281.
[8] N. Kishimoto, Y. Nakamura, M. Kato, H. Otsu, Effect of oxidation–reduction potential on an electrochemical Fenton-type process, Chemical Engineering Journal 260 (2015) 590–595.
[9] B. Sun, W. Gui, T. Wu, Y. Wang, C. Yang, An integrated prediction model of cobalt ion concentration based on oxidation-reduction potential, Hydrometallurgy 140 (2013) 102–110.
[10] P. Londhe, B. Patre, A.P. Tiwari, Design of single-input fuzzy logic controller for spatial control of advanced heavy water reactor, IEEE Transactions on Nuclear Science 61 (2) (2014) 901–911.
[11] R.M. Kamel, A. Chaouachi, K. Nagasaka, Retracted: wind power smoothing using fuzzy logic pitch controller and energy capacitor system for improvement micro-grid performance in islanding mode, 2010.
[12] X. Sun, K. Koh, B. Yu, M. Matsui, Fuzzy-logic-based v/f control of an induction motor for a DC grid power-leveling system using flywheel energy storage equipment, IEEE Transactions on Industrial Electronics 56 (8) (2009) 3161–3168.
[13] S. Won, C. Ra, Biological nitrogen removal with a real-time control strategy using moving slope changes of pH (mV)-and ORP-time profiles, Water Research 45 (1) (2011) 171–178.
[14] J.J. Carbajal-Hernández, L.P. Sánchez-Fernández, J.A. Carrasco-Ochoa, J.F. Martínez-Trinidad, Immediate water quality assessment in shrimp culture using fuzzy inference systems, Expert Systems with Applications 39 (12) (2012) 10571–10582.
[15] B. Zhang, C. Yang, H. Zhu, Y. Li, W. Gui, Evaluation strategy for the control of the copper removal process based on oxidation–reduction potential, Chemical Engineering Journal 284 (2016) 294–304.
[16] Y. Han, P. Shi, S. Chen, Bipolar-valued rough fuzzy set and its applications to the decision information system, IEEE Transactions on Fuzzy Systems 23 (6) (2015) 2358–2370.
[17] D.P. Pancho, J.M. Alonso, O. Cordón, A. Quirin, L. Magdalena, FINGRAMS: visual representations of fuzzy rule-based inference for expert analysis of comprehensibility, IEEE Transactions on Fuzzy Systems 21 (6) (2013) 1133–1149.
[18] R. Alcalá, M.J. Gacto, F. Herrera, A fast and scalable multiobjective genetic fuzzy system for linguistic fuzzy modeling in high-dimensional regression problems, IEEE Transactions on Fuzzy Systems 19 (4) (2011) 666–681.
[19] R. Alcalá, Y. Nojima, F. Herrera, H. Ishibuchi, Multiobjective genetic fuzzy rule selection of single granularity-based fuzzy classification rules and its interaction with the lateral tuning of membership functions, Soft Computing 15 (12) (2011) 2303–2318.
[20] C. Gao, Q. Ge, L. Jian, Rule extraction from fuzzy-based blast furnace SVM multiclassifier for decision-making, IEEE Transactions on Fuzzy Systems 22 (3) (2013) 586–596.
[21] B. Sun, W. Gui, Y. Wang, C. Yang, Intelligent optimal setting control of a cobalt removal process, Journal of Process Control 24 (5) (2014) 586–599.
[22] K. Nigam, A.K. McCallum, S. Thrun, T. Mitchell, Text classification from labeled and unlabeled documents using EM, Machine Learning 39 (2–3) (2000) 103–134.
[23] A. Daneshpazhouh, A. Sami, Semi-supervised outlier detection with only positive and unlabeled data based on fuzzy clustering, International Journal on Artificial Intelligence Tools 24 (03) (2015) 1550003.

[24] M. Claesen, F. De Smet, J.A. Suykens, B. De Moor, A robust ensemble approach to learn from positive and unlabeled data using SVM base models, Neurocomputing 160 (2015) 73–84.
[25] S. Jain, M. White, P. Radivojac, Estimating the class prior and posterior from noisy positives and unlabeled data, Advances in Neural Information Processing Systems (2016) 2693–2701.
[26] M. González, C. Bergmeir, I. Triguero, Y. Rodríguez, J.M. Benítez, On the stopping criteria for k-nearest neighbor in positive unlabeled time series classification problems, Information Sciences 328 (2016) 42–59.

Part III

Modeling and optimal control of the cobalt removal process

Chapter 6

Integrated modeling of the cobalt removal process

Contents

6.1 Process description and analysis

The cobalt removal process is the immediate subsequent step of the copper removal process. Among the impurities in the zinc sulfate solution, cobalt ions are the most detrimental one. Cobalt ions can significantly reduce the current efficiency in the electrowinning process, even when the concentration of cobalt ions is low. The overall performance of the purification process largely relies on the cobalt removal result. Similarly to copper removal, the cobalt removal process can be abstracted as a replacement reaction. However, there exists a high kinetic barrier in cobalt removal. Therefore, cobalt removal is extremely difficult.

The cobalt removal process can be expressed as follows:

$$Co^{2+} + Zn \rightarrow Zn^{2+} + Co. \tag{6.1}$$

In reaction (6.1), the standard potential of cobalt is -0.277 V, while the standard potential of zinc is -0.763 V. From the perspective of electrochemistry, cobalt ions with a more positive standard potential can be easily replaced by zinc with a more negative standard potential. However, in the practical cobalt removal process, the overvoltage phenomenon takes place during the deposition of cobalt ions from the solution, i.e., the deposition potential of cobalt ion is lower than its standard potential. In addition, the deposition potential increases when the temperature declines. The deposition potential of cobalt ion is around -0.4 V when the solution temperature is 80°C. The concentration of cobalt ions in the

Modeling, Optimization, and Control of Zinc Hydrometallurgical Purification Process
https://doi.org/10.1016/B978-0-12-819592-5.00017-X

neural leaching solution is relatively low, which indicates a low activity. Therefore, the practical deposition potential of cobalt ion is around −0.6 V, which is very close to the standard potential of zinc. The overvoltage phenomenon gives rise to the limited thermodynamic impetus of the replacement reaction between cobalt ions and zinc. However, hydrogen ions have a very low deposition overvoltage on cobalt, and the practical deposition potential of hydrogen ions is more positive than −0.6 V. As a result, the deposition potential of hydrogen ions on cobalt is more positive than the discharge potential of cobalt ions. The hydrogen ions will obtain electrons before the cobalt ions if the cobalt ions are deposited on the surface of zinc powder. In this situation, the replacement reaction is between zinc and hydrogen ions instead of zinc and cobalt ions. Due to the above reasons, the removal of cobalt ions is the most difficult one in the solution purification process [1]. To enable the removal of cobalt ion, the deposition potential of cobalt ions should be more positive than the deposition potential of hydrogen ions. If deep purification is needed, a relatively large difference between the deposition potential of zinc and cobalt is required to generate sufficient thermodynamic impetus for the replacement process. Cobalt has a more positive deposition potential on the surface of arsenic and antimony than hydrogen ions. In addition, the overvoltage of hydrogen ions on arsenic and antimony is much higher than that of cobalt. In practical production, salt of electropositive metals such as arsenic trioxide and antimonite is added in the zinc sulfate solution to catalyze cobalt removal by increasing the deposition potential of cobalt ions and suppressing the discharge of hydrogen ions [2].

The arsenic trioxide-activated cobalt removal process (ACP) is a widely used cobalt removal technology capable of deep purification. As shown in Fig. 6.1,

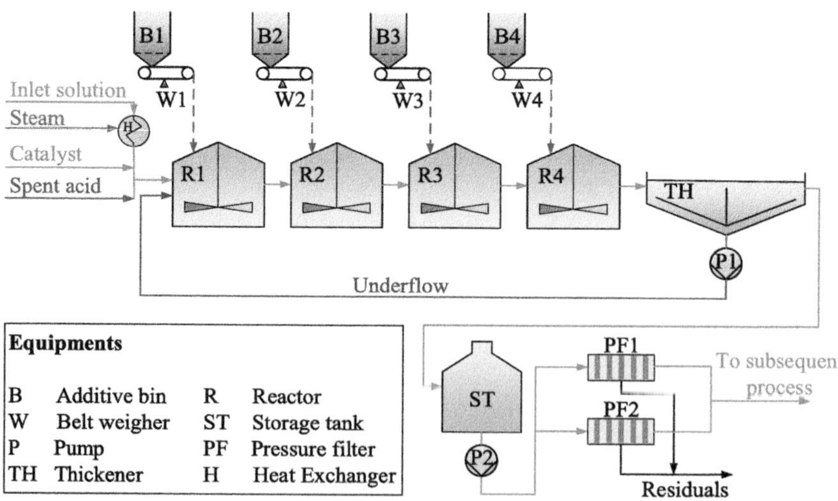

FIGURE 6.1 Arsenic trioxide-activated cobalt removal process.

ACP is mainly composed of several continuous stirred tank reactors (CSTRs) connected in series and a thickener. Zinc dust is added to each reactor to replace cobalt ions. Arsenic trioxide is added to the first reactor as catalyst. Spent acid is provided to create an acidic reaction environment. To guarantee that there exists sufficient reaction impetus, steam is used to increase the temperature of the inlet solution. Under the high temperature and acid reaction conditions, the following complex chemical and electrochemical reactions take place:

$$Cu^{2+} + Zn \rightarrow Zn^{2+} + Cu, \tag{6.2}$$

$$As^{3+} + 3Cu^{2+} + 4.5Zn \rightarrow Cu_3As + 4.5Zn^{2+}, \tag{6.3}$$

$$As^{3+} + Co^{2+} + 2.5Zn \rightarrow CoAs + 2.5Zn^{2+}, \tag{6.4}$$

$$As^{3+} + Ni^{2+} + 2.5Zn \rightarrow NiAs + 2.5Zn^{2+}, \tag{6.5}$$

$$2H^{+} + Zn \rightarrow Zn^{2+} + H_2 \uparrow, \tag{6.6}$$

where Zn is from the added zinc powder, As^{3+} is from arsenic trioxide, and Cu^{2+} is reserved copper ions after copper removal. Cobalt ions are gradually precipitated through forming a metal compound or alloy, e.g., CoAs and Cu_3As. After retention in consecutive reactors, the solution flows into the thickener in which liquid-solid separation takes place. As the reaction solution, the overflow of the thickener is sent to the subsequent cadmium removal process. However, crystal nuclei contained in the underflow benefit cobalt removal. Therefore, the underflow is recycled to the first reactor to promote cobalt cementation. The oxidation–reduction potential (ORP) is monitored online continuously.

In the operation of the cobalt removal process, the operator is mainly concerned with two important indicators, including one technical indicator and one economical indicator. The outlet cobalt ion concentration is the technical indicator, reflecting the performance of purification. The zinc powder dosage is the economical indicator, related to production cost. It is important to determine the zinc powder dosage of each reactor in the operation of ACP.

An excessive amount of additive is a waste of costly material, while an insufficient amount fails to remove the impurity adequately. The aim of operational optimization is to achieve the required purification performance using the minimal amount of zinc powder, while simultaneously keeping the stability of ACP. A human operator adjusts the manipulated variables by observing and analyzing the variation tendency of key process indicators. These indicators include:

- The flow rate of feeding zinc sulfate solution.
- The flow rate of recycled underflow.
- The flow rate of spent acid.
- The flow rate of arsenic trioxide.
- The dosage of zinc powder for each reactor.
- The temperature of each reactor.
- The ORP of each reactor.

However, ACP is highly nonlinear due to the inherent complexity of the reaction mechanism and frequent fluctuations of the inlet condition. The knowledge about the dynamic behavior of ACP is limited. In addition, there exist strong couplings among the process variables. Therefore, deciding the optimal combination of the manipulated variables is a challenging task for human operators. In some cases, a large variation of the inlet condition even causes failure to meet the technical requirement on outlet cobalt ion concentration.

6.2 Kinetics of cobalt removal reactions

In this section, we study the reaction mechanism of ACP from the perspective of chemical reaction kinetics, electrode reaction kinetics, and electrochemistry [3]. To start with, the influencing factors of the cobalt removal process are analyzed in Section 6.2.1. Then, the reaction type, the reaction steps, and the relation between ORP and the reaction rate are determined in Sections 6.2.2 and 6.2.3. In Section 6.2.4, a kinetic model of ACP is built based on a mechanistic study and verified via an experimental study.

6.2.1 Influencing factor analysis

ACP is an extremely complicated three-phase (gas-solid-liquid) reaction process with various influencing factors. The main factors affecting ACP are summarized as follows.

6.2.1.1 Temperature

In order to provide sufficient reaction impetus for the cobalt removal process, it needs to be performed in a high-temperature environment. Therefore, the reaction temperature is a very important control parameter for ACP. When the temperature is low, the cobalt removal performance is extremely poor due to a lack of sufficient reaction impetus. However, when the temperature is too high, the hydrogen ions in the solution are prone to be reduced, i.e., generating hydrogen. The high temperature also increases production costs and brings extra difficulty for operation, e.g., sampling. In the practical production process, the reaction temperature for ACP is generally controlled within the range of 75–85°C.

The influences of the temperature on ACP mainly include the following aspects.

Reaction rate

As the temperature increases, the cobalt removal process has a higher reaction speed. So increasing the temperature can increase the concentration diffusion rate of cobalt ions from the solution to the surface of the zinc powder particle. On the other hand, the overvoltage of hydrogen deposition will dramatically decrease when the temperature increases; however, the drop is much smaller than that of the cobalt deposition overvoltage. Therefore, increasing the temperature can enlarge the deposition potential difference between cobalt and zinc, which increases the reaction impetus. The promoting effects of increasing the temperature on cobalt removal can also be verified by the Arrhenius equation. From Eq. (6.7), it can be observed that the higher the temperature T, the larger the reaction rate k, which is more favorable for cobalt removal. We have

$$k = A_0 \exp(-\frac{E_a}{\mathrm{R}T}), \quad (6.7)$$

where A_0 is a preexponential factor or frequency factor, E_a is the activation energy, R is the ideal gas constant, and T is the reaction temperature.

Reaction product morphology

Under different temperature conditions, the cobalt removal rates are different, resulting in different cobalt content in the reaction product. When the temperature is low, the reaction rate of cobalt removal is low. The cobalt content in the reaction product is low. So cobalt has a small influence on the morphology of the reaction product, which is a hexagonal lattice. When the temperature is high, the reaction rate of cobalt removal is high. The cobalt content in the reaction product is high. The mismatch between the lattices of zinc and cobalt is more obvious. Cobalt has a large influence on the morphology of the reaction product. The reaction product is a nodular cobalt-zinc alloy.

Distribution of cathode current

Under different temperature conditions, the distributions of cathode current are different. Van der Pas and Dreisinger found through experimental study that: (i) at 85°C, 50% of the cathode current is used to reduce cobalt; and (ii) at 90°C, the reduction of hydrogen ions becomes the most important reaction, and only 2% of the cathode current is used for the reduction of cobalt. The most possible cause of this change in the current distribution lies in the change of reaction product composition. At 85°C, the main content of the cobalt-zinc alloy is zinc. So the hydrogen reduction on the cathode is mainly controlled by zinc. Due to the high overvoltage of hydrogen on zinc, the reduction of hydrogen ions is inhibited. At 90°C, the main content of the cobalt-zinc alloy is cobalt. So the hydrogen reduction on the cathode is mainly controlled by cobalt. As cobalt can catalyze the reduction of hydrogen ions, a large number of hydrogen ions are reduced to hydrogen, causing the waste of zinc powder.

6.2.1.2 Dosage of arsenic trioxide

Arsenic trioxide is a catalyst of the cobalt removal process. If the dosage of arsenic trioxide is insufficient, high cobalt removal performance cannot be guaranteed. Therefore, a sufficient amount of arsenic trioxide must be provided to achieve the objective of cobalt removal. The influence of arsenic trioxide on the cobalt removal performance can also be observed from the following chemical reactions [5]:

$$As_2O_3 + 2NaOH \rightarrow 2NaAsO_2 + H_2O, \tag{6.8}$$

$$AsO_2^- + H^+ \rightarrow HAsO_2, \tag{6.9}$$

$$Cu^{2+} + Zn \rightarrow Zn^{2+} + Cu, \tag{6.10}$$

$$HAsO_2 + 3Cu^{2+} + 3H^+ + 4.5Zn \rightarrow Cu_3As + 4.5Zn^{2+} + 2H_2O, \tag{6.11}$$

$$HAsO_2 + Co^{2+} + 3H^+ + 2.5Zn \rightarrow CoAs + 2.5Zn^{2+} + 2H_2O, \tag{6.12}$$

$$HAsO_2 + Ni^{2+} + 3H^+ + 2.5Zn \rightarrow NiAs + 2.5Zn^{2+} + 2H_2O. \tag{6.13}$$

The first two reactions describe the preparation process of arsenic trioxide. As arsenic trioxide can hardly dissolve in zinc sulfate solution, it is first treated using sodium hydroxide solution to be converted to sodium metaarsenite. Sodium metaarsenite solution has a strong basicity. It will hydrolyze and produce subsulfate, which can attach to the surface of zinc powder particles and therefore hinder cobalt removal. So it is then acidized to metaarsenous acid by adding spent acid. The last four reactions are the main reactions of the cobalt removal process, according to which the chemometric relation between arsenic trioxide and impurities (Cu^{2+}, Co^{2+}, Ni^{2+}) can be derived. This indicates that the dosage of arsenic trioxide must be sufficient. Otherwise, the removal of cobalt ions is insufficient. However, if the dosage of arsenic trioxide is excessive, the toxic arsine may be generated. We have

$$HAsO_2 + 3Zn + 6H^+ \rightarrow AsH_3 \uparrow + 3Zn^{2+} + 2H_2O. \tag{6.14}$$

6.2.1.3 Dosage of zinc powder

As a participant in cobalt removal, the dosage of zinc powder directly affects the performance of cobalt removal. If the dosage of zinc powder to replace cobalt is insufficient, then the cobalt ion concentration is not properly reduced. If excessive zinc powder is added, excess zinc powder will react with hydrogen ions in the solution, generating hydrogen and causing an increase in the pH. If arsenic trioxide is also excessive, then elemental arsenic will be generated. If zinc powder is overly excessive, it is prone to produce toxic arsine, as introduced above. We have the following reactions:

$$Zn + 2H^+ \rightarrow Zn^{2+} + H_2 \uparrow, \tag{6.15}$$

$$HAsO_2 + 1.5Zn + 3H^+ \rightarrow As + 1.5Zn^{2+} + 2H_2O. \tag{6.16}$$

6.2.1.4 Flow rate of spent acid

The flow rate of spent acid affects the pH value of the solution inside the reactor. Experimental results indicate that [4.0, 4.4] is the optimal range of pH. If spent acid is insufficient, then the pH value is too high, and the basic zinc sulfate or zinc hydroxide will be produced and deposited on the surface of zinc powder particles. This could hinder cobalt removal. If spent acid is excessive and the pH value is too low, then the deposition potential of hydrogen ions will increase. Zinc powder will directly react with hydrogen ions, which is a waste of zinc powder and also not beneficial for cobalt removal. However, compared with a higher pH, a lower pH is more harmful to the cobalt removal process, because the reduction of hydrogen ions not only consumes zinc powder, but also blocks the substrate where the cobalt removal reaction takes place. Therefore, the flow rate of spent acid is a compromise which minimizes the adverse effects of both situations.

6.2.1.5 *Concentration of zinc ions and copper ions*

If the concentration of zinc ions in the feeding solution is too high, then the concentration of zinc ions surrounding zinc powder particles is too high. The difficulty of zinc ions entering the solution is increased. As a result, zinc can hardly lose electrons and be converted to zinc ions. In addition, it is difficult for cobalt ions to obtain electrons and deposit on the surface of zinc powder particles. Cobalt ions can be quickly removed to a low concentration if zinc ions are not present in the solution.

An appropriate amount of copper ions in the feeding solution is beneficial for cobalt removal. Compared with cobalt removal activated by antimonic salt, in ACP, copper ions have a larger promoting effect on cobalt removal. Therefore, in practice, the copper ions are not completely removed in the copper removal process. Part of the copper ions are reserved for the cobalt removal process. However, excessive copper ions will increase zinc powder consumption. The promoting effect of copper ions on cobalt removal can be explained from two aspects:

(i) Copper ions react with metaarsenous acid, producing copper arsenide. Copper arsenide will form a microcell with zinc powder, which acts as a cathode to promote the precipitation of cobalt ions.
(ii) Although the formation of zinc-cobalt alloy will reduce the potential difference between cobalt and zinc, thereby reducing the thermodynamic impetus of the replacement reaction, if the formation of the alloy is not between the replacement metal and the replaced metal, but two or more replaced metals, then the situation is totally different. In this case, the metal with the most positive deposition potential (such as copper) can move the deposition potential of some electronegative impurities towards a positive direction. As a result, the potential difference of the original zinc-cobalt microcell is increased, i.e., the thermodynamic impetus of the replacement reaction is increased.

6.2.1.6 *Other influencing factors*

In addition to the above factors, there exist other factors influencing the cobalt removal process, e.g., the concentration of impurity ions such as cadmium ions in the feeding solution, the flow rate of the feeding solution, and the size of zinc powder particles. Cadmium ions have little effect on the rate of cobalt removal; however, the synergistic effects of cadmium ions and arsenic/copper can improve cobalt removal performance. The improvement happens only when the cadmium ion concentration is low. Increasing the cadmium ion concentration cannot further improve cobalt removal [6]. Chloride ions can catalyze cobalt removal, and the catalytic effect increases with increasing chloride ion concentrations. However, due to the limitation on the maximum chloride ion concentration imposed by the electrowinning process, the chloride ion concentration cannot be increased without limitation. The flow rate of feeding solution determines the average residence time of the solution in the reactor. If other reaction conditions are kept constant, a larger flow rate indicates a shorter residence time. So the reaction rate must be increased to avoid the sudden increase of outlet cobalt ion concentration. If the flow rate of feeding solution is low, the residence time is long. In this case, the reaction rate can be appropriately decreased to reduce zinc powder consumption. The rate of cobalt removal increases as the size of the zinc powder particles decreases. However, if the particle size is too small, redissolution of cobalt ions may take place. The flow rate and solid content of the underflow affect the amount of reaction substrate in the reactors. In practice, the fluctuations of these two variables are small in most cases.

As the feeding solution of ACP contains not only cobalt ions, but also other ions, e.g., copper ions, besides the main reactions, there exist other side reactions taking place simultaneously. These reactions interact with each other. In addition, the numerous influencing factors affect ACP from different perspectives, leading to the complex dynamics of ACP.

6.2.2 Analysis of reaction type and steps

In ACP, the replacement between cobalt ions and zinc powder is enabled by arsenic trioxide. Therefore, the function of arsenic trioxide, reaction steps, and reaction type have been the main research focus.

There exist two theories on the mechanism by which arsenic trioxide promotes cobalt removal, i.e., the "alloy theory" and the "substrate theory." The "alloy theory" claims that cobalt ions are removed by forming inert alloy with the catalyst via reaction (6.4). Fountoulakis found that the function of copper ions and metal salt is to form a stable alloy with cobalt, which reduces the activity of cobalt [7]. Polcaro et al. concluded that the formation of metal compounds gives rise to a larger potential difference between cobalt and zinc, resulting in larger reaction impetus [8]. The "substrate theory" explains that the reaction products of reactions (6.3) and (6.4) serve as a substrate for the cementation of cobalt. On the surface of the substrate, cobalt ions have a relatively low overvoltage, while the hydrogen ions have a relatively high overvoltage. Therefore, cobalt ions can obtain electrons and deposit on the substrate. Van der Pas and Dreisinger studied the activation of cobalt removal by antimonate [1]. It was found that in the early stage of the reaction, the replacement reactions between copper, antimony, and zinc first take place. During this stage, cobalt still exists as ions in the solution. Copper and antimony take up electrons and deposit on the surface of zinc powder. Then, the deposited copper and antimony act as the substrate for cobalt ions [1]. Fugleberg et al. studied the ACP. It was observed that arsenic ions were deposited before the cementation of cobalt ions [5], which is similar to Van der Pas and Dreisinger's results.

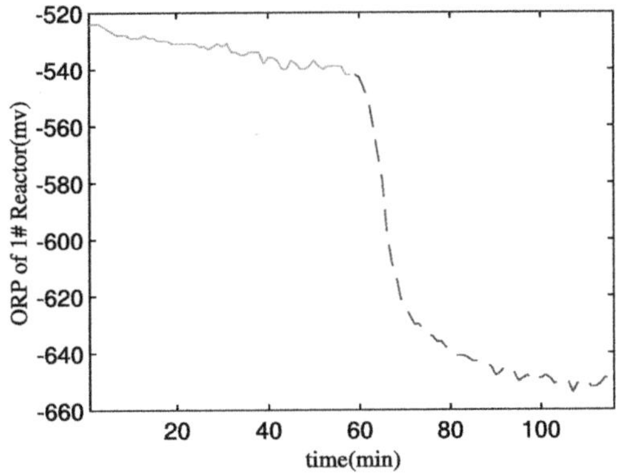

FIGURE 6.2 ORP variation of reactor 1.

The "alloy theory" and "substrate theory" explain the reaction steps and the function of arsenic trioxide from different perspectives. In order to verify the reaction type and steps of ACP, we studied the production data collected during the interval between shut-down and start-up. As shown in Fig. 6.2, in the beginning stage of the reaction, ORP changes very slowly, and the shape of the ORP trajectory is similar to a titration curve. After a certain period, ORP declines dramatically and then stabilizes around a certain value. As the reaction in ACP is essentially an oxidation–reduction reaction, the different variation trends indicate two reaction stages with different characteristics. In addition, it was noticed that copper ions were totally removed in the beginning stage of the reaction, which is in agreement with Van der Pas and Dreisinger's results.

Based on the analysis, the reaction steps of ACP can be concluded as follows.

(i) In the beginning stage of the reaction, copper ions, arsenic trioxide, and zinc powder react and produce a copper-arsenic alloy on the surface of the zinc powder particles. The alloy then acts as a substrate for cobalt cementation. We have the following reaction:

$$HAsO_2 + 3Cu^{2+} + 3H^+ + 4.5Zn \rightarrow Cu_3As + 4.5Zn^{2+} + 2H_2O. \quad (6.17)$$

(ii) Then, the substrate forms a microcell with zinc powder. Cobalt ions take up electrons from zinc powder and are cemented on the substrate surface. With the proceeding of the reaction, cobalt, arsenic, and copper form a stable alloy, which reduces the activity of cobalt. Therefore, the cobalt ion concentration declines in the zinc sulfate solution. This stage can be expressed as the following reaction:

$$Cu_3As + Co^{2+} + Zn \rightarrow 3Cu + CoAs + Zn^{2+}. \quad (6.18)$$

In ACP, the function of arsenic trioxide is twofold. On the one hand, it reacts with copper ions and zinc powder to produce a copper-arsenic alloy beneficial for cobalt cementation. On the other hand, as cobalt can be easily deposited on the surface of arsenic, the copper-arsenic alloy forms a microcell with zinc powder. Cobalt ions take up electrons and are deposited on the surface of arsenic. The deposited cobalt forms an alloy with arsenic. The existence of copper enlarges the potential difference between cobalt and zinc, which increases the thermodynamic impetus of the reaction. Based on the above analysis, it is concluded that the cobalt cementation takes place on the surface of a substrate composed of CoAs and another metal alloy, and the ACP consists of two steps.

To investigate the reaction type of ACP, Polcaro et al. studied the cobalt removal process with copper ions and antimonate as catalyst [8]. In the investigation, the deposition of copper and antimony was first studied in the solution without cobalt ions, and the result indicates that the deposition of copper and antimony is a first-order reaction controlled by the diffusion process. Then, as a comparison, the deposition of cobalt was studied. The reaction rate of cobalt removal significantly accelerated when the stirring rate increased from 220 rpm to 300 rpm. The result indicates that similarly to copper and antimony removal, the reaction rate of cobalt removal is largely influenced by the diffusion process. In Did's experiment, when the stirring rate increased from 500 rpm to 4000 rpm, there was no obvious change in the reaction rate of cobalt removal, which demonstrates that the reaction rate of cobalt removal is controlled by the chemical reaction [4]. Fugleberg et al. estimated the activation energy of ACP using the Arrhenius equation. An activation energy of 70 KJ/mol means that ACP is controlled by a chemical reaction rather than a diffusion process [5].

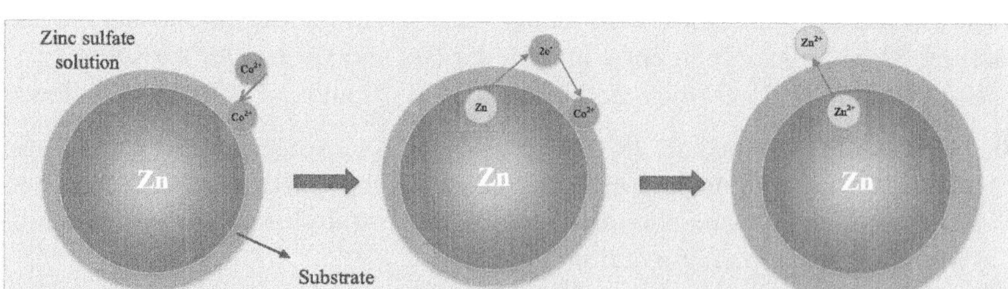

FIGURE 6.3 Microscopic reaction step in the cobalt removal process.

As shown in Fig. 6.3, from a microscopic point of view, the reaction in ACP can be divided into the following steps:

Step 1: Diffusion of cobalt ions from the solution to the substrate surface.
Step 2: Cobalt ions take up electrons and are deposited on the surface of the substrate or the cathode of the microcell.
Step 3: Zinc loses the electron and is oxidized to zinc ions.
Step 4: Diffusion of zinc ions to the solution.

The reaction step with the lowest rate is the controlling step of the reaction. For **Step 3**, if the anode of the microcell is active and the activity of zinc ions in the solution is not high, then it will not become the controlling step. For **Step 1** and **Step 4**, their progressing rates are relatively low and become the controlling step for many replacement reactions, e.g., copper removal and antimony removal. However, for the reaction with difficulties in the discharge of ions, **Step 2** becomes the controlling step, e.g., ACP.

In fact, a chemical reaction belongs to a different reaction type under different reaction conditions. For the reaction in ACP, when the stirring rate is low, an increase in the stirring rate can promote cobalt removal. However, when the stirring rate is higher than 500 rpm, an increase in the stirring rate has little effect on the reaction rate of cobalt removal. Therefore, both diffusion processes and the chemical reaction can affect the rate of cobalt removal. However, the influence of the diffusion process is limited. Essentially, cobalt removal is controlled by the chemical reaction.

6.2.3 Relation between ORP and reaction rate

In ACP, the coexistence of various oxidation–reduction reactions leads to a complex electrode reaction system.

At the anode, we have

$$Zn \rightarrow Zn^{2+} + 2e^-, \qquad E^0 = -0.763\ V, \tag{6.19}$$

and at the cathode, we have

$$Cu^{2+} + 2e^- \rightarrow Cu, \qquad E^0 = 0.337\ V, \tag{6.20}$$

$$Co^{2+} + 2e^- \rightarrow Co, \qquad E^0 = -0.277\ V, \tag{6.21}$$

$$Ni^{2+} + 2e^- \rightarrow Ni, \qquad E^0 = -0.250\ V, \tag{6.22}$$

$$2H^+ + 2e^- \rightarrow H_2 \uparrow, \qquad E^0 = 0\ V. \tag{6.23}$$

According to the independence theory of parallel electrode reactions, each electrode reaction is independent of the remaining electrode reactions. The only common point of the electrode reactions is the electrode potential, or mixed potential, which is also known as ORP. The reaction rate of electrode reaction relies on the rate of ion discharging, which is determined by the ORP [9][10]. The ORP is not the equilibrium potential. The value of ORP lies between the equilibrium potentials of the anode and cathode reactions. It is a synthetical

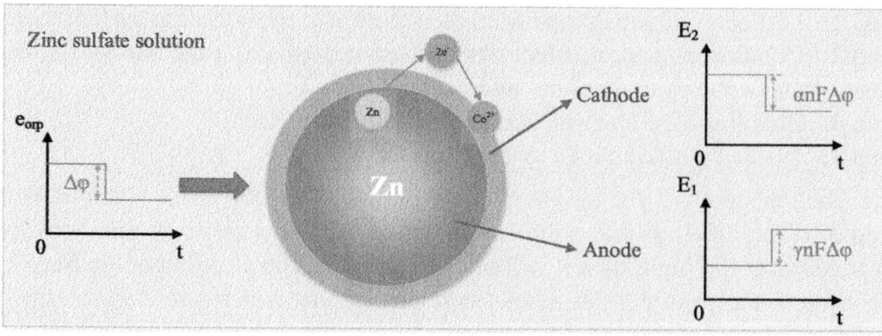

FIGURE 6.4 The effect of electrode potential change on anode and cathode activation energy.

result of the electrode reactions. As ORP is not equal to their equilibrium potentials, both anode and cathode reactions proceed continuously.

The ORP can affect the rate of the electrode reaction by changing its activation energy. Activation energy is the energy a molecule requires when transferring from the normal state to the active state ready for chemical reaction. Consider the following reduction reaction:

$$\mathrm{O} + \mathrm{ne}^- \rightarrow \mathrm{R}, \tag{6.24}$$

where n is the number of electrons O obtains in the reduction reaction. From the perspective of electrode reaction kinetics, a positive charge of nF is transferred from the solution to the electrode with the reaction of 1 mol O. As shown in Fig. 6.4, if the electrode potential deceases by $\Delta\varphi$, then the total potential energy will decrease by $nF\Delta\varphi$. The activation energy of the anode and cathode will increase or decrease by certain fractions of $nF\Delta\varphi$, i.e.,

$$E_1' = E_1 + \gamma nF\Delta\varphi, \tag{6.25}$$

$$E_2' = E_2 - \alpha nF\Delta\varphi, \tag{6.26}$$

where γ and α represent the influence of the electrode potential on the activation energy, or the transfer coefficient of the anode and cathode reactions, and satisfy

$$\gamma + \alpha = 1. \tag{6.27}$$

If we use E_e and e_{eq} to denote the activation energy and equilibrium potential of the cathode reaction, then the reaction rate can be expressed as

$$k = A_0 \exp\left(-\frac{E_e + 2\alpha \mathrm{F}(e_{\mathrm{orp}} - e_{\mathrm{eq}})}{\mathrm{R}T_c}\right), \tag{6.28}$$

where e_{orp} is ORP.

According to (6.28), a more negative ORP will result in a higher reaction rate, and vice versa. This relation is also observed from production data. ACP is the controlled chemical reaction and happens on the surface of the substrate. Assume the solution in the reactor is perfectly mixed. Then by applying the mass balance principle, the dynamics of the cobalt removal reaction in each reactor can be formulated as

$$\frac{dc}{dt} = \frac{f_{\text{in}}}{V} c_{\text{in}} - \frac{f_{\text{out}}}{V} c - kAc, \tag{6.29}$$

where V is the volume of the reactor, f_{in} and f_{out} are the inlet and outlet flow rate of the reactor, c_{in} and c are the inlet and outlet cobalt ion concentrations of the reactor, k is the reaction rate, and A is the surface area of the substrate or seed crystal in unit volume. As shown in Figs. 6.5, 6.6, and 6.7, a quasilinear relation between kA and ORP was observed under different flow rates. (Before the identification of the model parameters, the value of A cannot be obtained. Therefore, kA was treated as unity.) When the ORP is more negative, the value of kA is larger. As the value of A stays in a certain range during a certain time interval, a larger kA indicates a larger k.

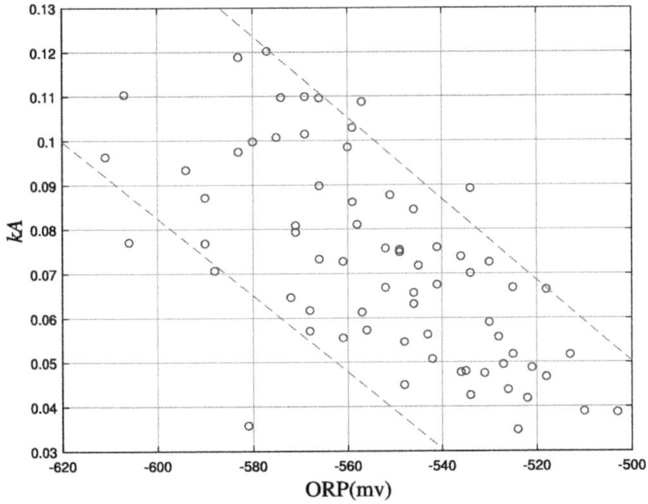

FIGURE 6.5 The relation between ORP and kA when the inlet flow rate is between 150 and 200 m^3.

6.2.4 Kinetic model construction

According to the results in Section 6.2.3, the dynamics of ACP can be described by

$$\frac{dc}{dt} = \frac{f_{\text{in}}}{V} c_{\text{in}} - \frac{f_{\text{out}}}{V} c - kAc. \tag{6.30}$$

Among the variables, V is constant, f_{in} and f_{out} are not controlled by the cobalt removal process, and c_{in} is determined by the feeding conditions and the opera-

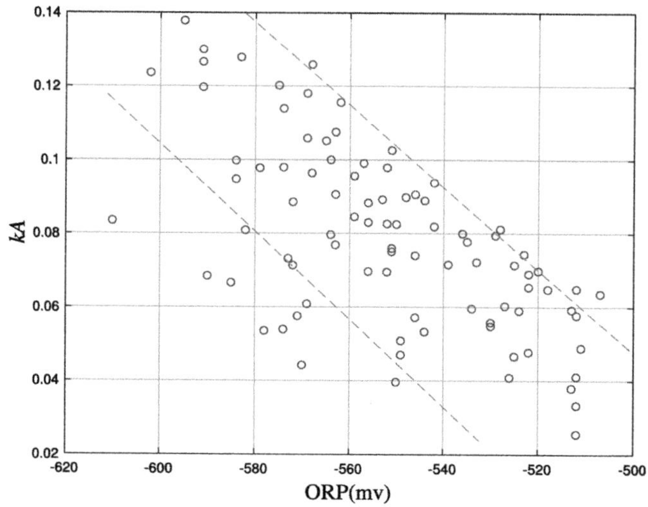

FIGURE 6.6 The relation between ORP and kA when the inlet flow rate is between 200 and 250 m^3.

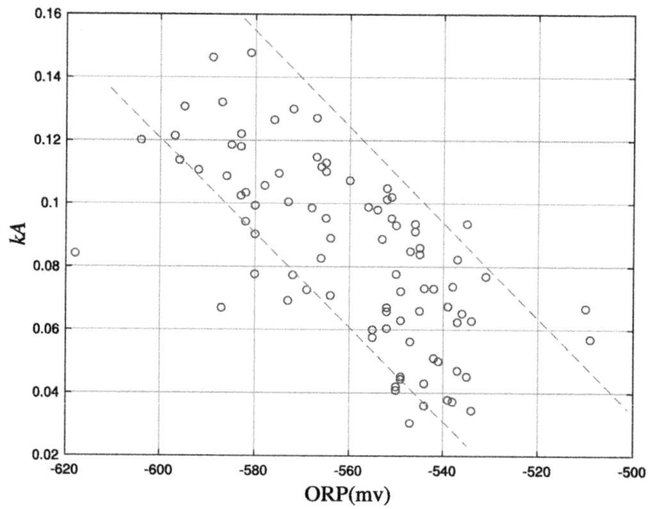

FIGURE 6.7 The relation between ORP and kA when the inlet flow rate is between 250 and 300 m^3.

tions conducted in the upstream processes. Therefore, in ACP, the reaction rate can be controlled by adjusting the values of k and A.

Substituting Eq. (6.28) into Eq. (6.30),

$$\frac{dc}{dt} = \frac{f_{\text{in}}}{V}c_{\text{in}} - \frac{f_{\text{out}}}{V}c - A_0\exp(-\frac{E_e + 2\alpha \mathrm{F}(e_{\text{orp}} - e_{\text{eq}})}{\mathrm{R}T_c})Ac. \tag{6.31}$$

In Eq. (6.31), the value of A is determined by the amount of substrate, e.g., cobalt-arsenic alloy. As shown in Fig. 6.1, the solid in a reactor consists of two parts: the seed crystal in the underflow from the thickener g_s and the seed crystal produced in real-time during the reaction g_r.

From the viewpoint of mass balance, the mass of the substrate in reactor #1 holds following the relation with the solid mass in the underflow:

$$g_s = \frac{f_{uf}}{f_0 + f_{uf}} g_{uf}, \tag{6.32}$$

where g_s is the mass of substrate in unit volume of the reactor, g_{uf} is the mass of substrate in unit volume of the reactor, and f_0 and f_{uf} are the flow rates of the inlet solution and underflow. According to the chemical assay, the value of g_s is 20 to 60 times as high as the value of g_r. The seed crystal produced in real-time is negligible compared with the seed crystal in the underflow. In addition, the value of g_r is hard to obtain. Therefore, the substrate mass in unit volume of the reactor is usually considered as g_s.

Assume that the shape of the substrate particle is spherical, the radius, volume, and surface area of each particle are r, V_0, and A_{s0}, the total volume of the particles in the reactor is V_s, the density of the solid content in the reactor is ρ, and V_s, V_0, and A_{s0} satisfy [8]

$$\rho V_s = g_s V, \tag{6.33}$$

$$V_0 = \frac{4}{3}\pi r^3, \tag{6.34}$$

$$A_{s0} = 4\pi r^2. \tag{6.35}$$

Then the total surface area of the particles in the reactor is

$$A_0 = \frac{V_s}{V_0} A_{s0} = g_s \frac{3V}{\rho r}, \tag{6.36}$$

where g_s can be obtained using (6.32) and ρ and r are determined by the physical properties of the particles.

If we denote $\frac{3}{\rho r}$ as β_u, then

$$A_0 = \beta_u g_s V. \tag{6.37}$$

Therefore, the surface area of the substrate in unit volume of the reactor is

$$A_u = \beta_u g_s. \tag{6.38}$$

In practice, the reaction rate cannot increase with A_u unlimitedly. In addition, the surface area of the substrate is not fully effective. If the surface is attached by basic zinc sulfate or if the solution is viscous due to a very high zinc ion concentration, the ACP is hindered and the effective surface area for cobalt deposition is shrunk. Therefore, the effective surface area per unit volume of the reactor is

$$A = \mu_u \beta_u g_s, \tag{6.39}$$

where μ_u is the ratio of the effective surface area.

If we denote $\beta_u g_s$ and μ_u as A_g and β, respectively, then

$$\frac{dc}{dt} = \frac{f_{in}}{V} c_{in} - \frac{f_{out}}{V} c - A_0 \exp(-\frac{E_e + 2\alpha F(e_{orp} - e_{eq})}{RT_c}) \beta A_g c. \tag{6.40}$$

The physical meanings and units of the model parameters are shown in Table 6.1.

In Eq. (6.40), A_0, E_e, α, e_{mix}, β, and A_g are unknown and need to be identified. Among these parameters, A_0, β, and A_g are correlated; A_0 represents the effective collision frequency between cobalt ions and substrate, and β and A_g are the ratio of effective surface area and reaction surface area of the substrate, respectively. These three parameters are physically closely related. They are all related to the contact between cobalt ions and substrate. In addition, their ranges are difficult to determine. Therefore, A_0, β, and A_g are combined into one parameter, $A_\beta = A_0 \beta A_g$. Eq. (6.40) can be further rewritten as

$$\frac{dc}{dt} = \frac{f_{in}}{V} c_{in} - \frac{f_{out}}{V} c - A_\beta \exp(-\frac{E_e + 2\alpha F(e_{orp} - e_{eq})}{RT_c}) c, \tag{6.41}$$

where A_β, E_e, α, and e_{eq} are model parameters to be identified.

6.3 First-principle/machine learning integrated process modeling

In Section 6.2.4, the kinetic model was built. The advantage of the kinetic model is that it can describe the underlying physicochemical laws of a process. However, as discussed in Section 6.2, the dynamics of ACP are influenced by various factors which cannot be explicitly covered by the kinetic model. As shown in Fig. 6.8, the consumption rate of Co^{2+} is determined by the main reactions [3]. However, the main reactions interact with side reactions. In addition, the interactions between ACP and other unit processes give rise to the fluctuations of inlet and reaction conditions, which then change the proportions and interaction manner of the main and side reactions. As a result, the ACP exhibits various working conditions with different dynamics.

6.3.1 Integrated modeling framework

In order to comprehensively describe the dynamics of ACP, a CSS-based integrated modeling framework has been proposed (Fig. 6.9) [11]. In order to make use of the advantages of the first-principle model (FPM) and the data-driven

TABLE 6.1 Parameters in the kinetic model.

Parameter	Physical meaning	Unit	Value
c_{in}	Inlet cobalt ion concentration	mg/L	
c	Outlet cobalt ion concentration	mg/L	
V	Volume of reactor	m^3	
g_s	Precipitant content in unit volume of the reactor	kg/m^3	
β	The ratio of effective reaction surface area	–	
A_0	Frequency factor of cobalt removal reaction	1/s	
R	Ideal gas constant	$J/(mol \cdot K)$	8.314472
T_c	Reaction temperature	K	
F	Faraday constant	C/mol	96485
E_e	Standard activation energy of the cobalt removal reaction	kJ/mol	
e_{orp}	Oxidation–reduction potential of the solution	V	
e_{eq}	Equilibrium potential of cobalt ions	V	
α	Transfer coefficient of electrode potential change to cathode activation energy	–	
A_g	Surface area of the substrate in unit volume of the reactor	1/s	

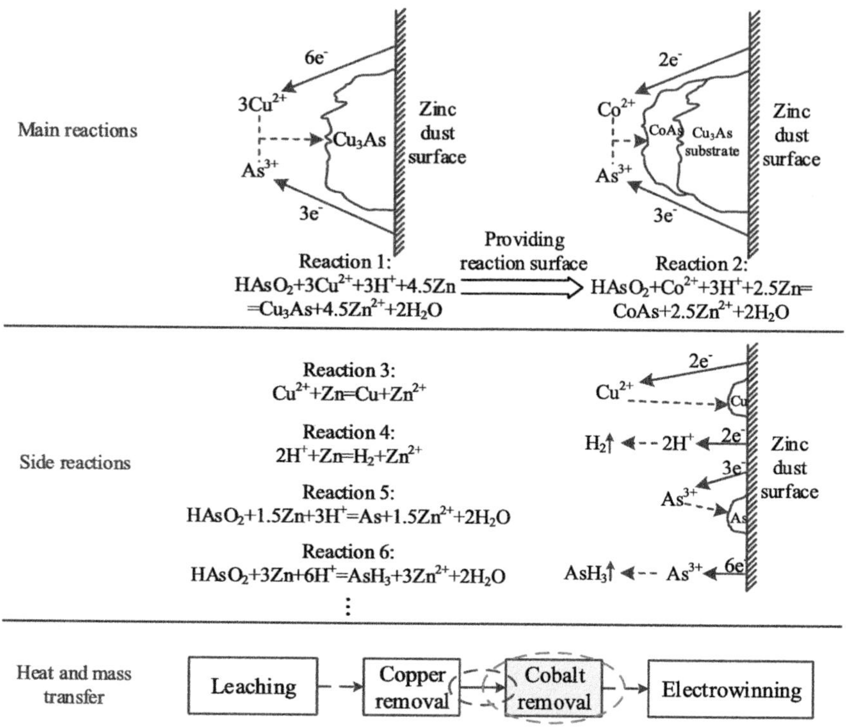

FIGURE 6.8 Main and side reactions in the cobalt removal process.

model, the process model is formulated in an integrated form, which is a combination of a FPM and a machine learning-based input/output model (ML-IOM). The FPM is considered as the nominal kinetic model. The ML-IOM accounts for the deviation between the output of the nominal kinetic model and the real value of the concerned index. Considering that ACP exhibits multiple working conditions, the deep features are first extracted to divide the comprehensive state space into different subspaces. Each subspace is associated with one type of working condition. Then, the submodels under each working condition are obtained by identifying and training the model parameters, i.e., the FPM and ML-IOM. Therefore, the dynamic process model is formulated as a weighted sum of the submodels, which has the same formulation as (2.15) in Chapter 2:

$$\mathbf{x}_O(k+1)=\mathbf{x}_O(k)+\sum_{i=1}^{N_p} P(i|\mathbf{z})[\mathbf{f}_k^{(i)}(\mathbf{x}_O(k),\Theta(k))+\mathbf{U}^{(i)}(\mathbf{z}(k))], \qquad (6.42)$$

where i indicates the ith working condition, $\mathbf{f}_k^{(i)}(\mathbf{x}_O(k),\Theta(k))$ can be derived from the nominal kinetic model or the FPM, $\mathbf{U}^{(i)}(\mathbf{z}(k))$ is the machine learning-based data compensation model, and $\mathbf{z}$ are the latent variables. The weights

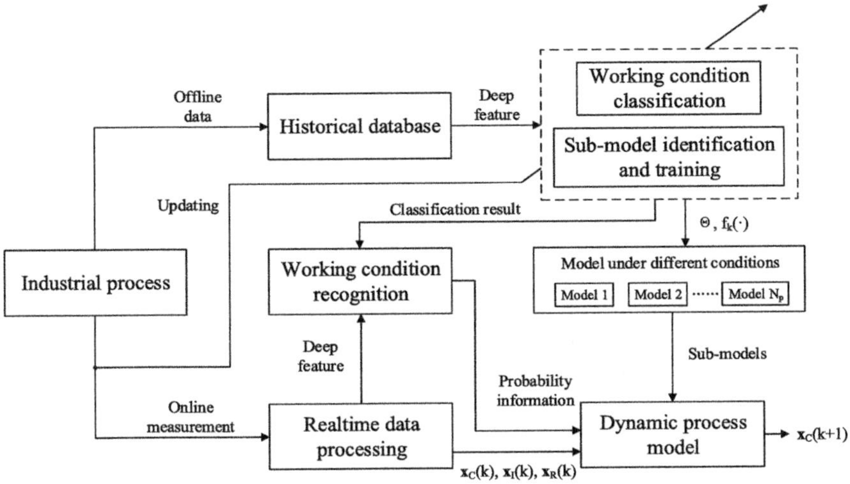

FIGURE 6.9 CSS-based integrated modeling framework of ACP.

$P(i|\mathbf{z})$ are the probabilities that the current process belongs to each working condition. It updates continuously with the evolution of ACP in the comprehensive state space.

6.3.2 Working condition classification

In the CSS modeling framework, parameters of the FPM and the machine learning model under each working condition are different. To automatically classify the working condition, the deep features of ACP are first extracted. Then, the comprehensive state space is divided into subspaces using deep features. Each of the subspaces corresponds to a working condition.

6.3.2.1 *Deep feature extraction*

In order to decrease the computation load, the high-dimensional raw data have to be mapped into low-dimensional latent variables, which can be regarded as features of raw data. The mapping can be achieved using deep feature extraction (DFE), which can recognize the patterns of raw data. The advantage of DFE lies in its ability to learn very complex nonlinear functions. This is achieved using multiple levels of representation [12][13][14].

Stacked auto-encoder (SAE) is a typical unsupervised multilevel deep learning network [15]. In SAE, multiple auto-encoders are stacked to form a multilevel structure. In each auto-encoder, there exist one input layer, one output layer, and one hidden layer (Fig. 6.10). It first encodes the inputs to latent variables, and then the latent variables are used to reconstruct the inputs in the output

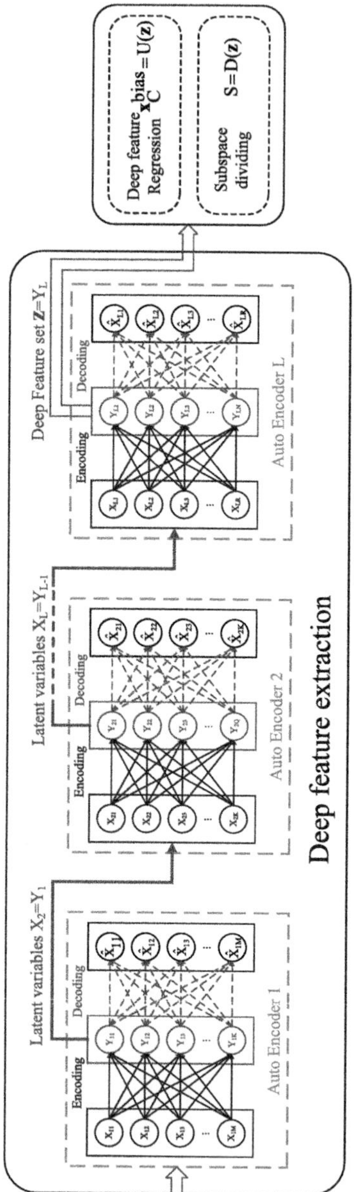

FIGURE 6.10 Deep feature extraction, regression, and subspace division.

layer. For the ith auto-encoder,

$$\mathbf{Y}_i = \mathbf{H}(\mathbf{W}_i \mathbf{X}_i + \mathbf{b}_i), \tag{6.43}$$

$$\hat{\mathbf{Y}}_i = \mathbf{H}'(\mathbf{W}_i' \mathbf{Y}_i + \mathbf{b}_i'), \tag{6.44}$$

where $\mathbf{X}_i \in \mathbb{R}^{n_i}$ is the input of an auto-encoder, $\mathbf{Y}_i \in \mathbb{R}^{m_i}$ is the latent variables to be extracted, $\hat{\mathbf{Y}}_i \in \mathbb{R}^{n_i}$ is the reconstructed outputs, $\mathbf{H}(\cdot)$ and $\mathbf{H}'(\cdot)$ are activation functions, e.g., sigmoid function, and $\mathbf{W}_i \in \mathbb{R}^{m_i \times n_i}$, $\mathbf{W}_i' \in \mathbb{R}^{n_i \times m_i}$, $\mathbf{b}_i \in \mathbb{R}^{m_i}$, and $\mathbf{b}_i' \in \mathbb{R}^{n_i}$ are weighting matrices and bias vectors, which are the design parameters obtained by minimizing the reconstruction error on a training set with M samples:

$$\ell_{\text{SAE}}(\mathbf{W}_i, \mathbf{W}_i', \mathbf{b}_i, \mathbf{b}_i') = \sum_{j=1}^{M} \|\mathbf{X}_i^{(j)} - \hat{\mathbf{Y}}_i^{(j)}\|^2. \tag{6.45}$$

The extracted latent variables can be regarded as features of the inputs, which are then used as inputs of a subsequent auto-encoder. The training of SAE includes a pretraining and a fine-tuning stage. In the pretraining stage, the auto-encoders are trained layer by layer. After the pretraining of each auto-encoder, the latent variables of the last auto-encoder are outputted as the deep features $\mathbf{z} = \mathbf{Y}_{\text{L}}$. Then, the regression function $U(\mathbf{z})$ is pretrained by minimizing the regression error:

$$\ell_{\text{Regression}}(\boldsymbol{\alpha}_1, \cdots, \boldsymbol{\alpha}_N) = \sum_{j=1}^{M} \|\mathbf{x}_{\text{O}}^{\text{deviation}^{(j)}} - \mathbf{U}(\mathbf{z}^{(j)})\|^2, \tag{6.46}$$

where $\mathbf{x}_{\text{O}}^{\text{deviation}^{(j)}}$ and $z_i^{(j)}$ are the FPM deviation and the deep features of the jth training sample, and $\mathbf{U}(\cdot)$ is a regression function of the deep feature set $\mathbf{z}$. After pretraining, $[\boldsymbol{\alpha}_1, \cdots, \boldsymbol{\alpha}_N]$ and $[\mathbf{W}_i, \mathbf{W}_i', \mathbf{b}_i, \mathbf{b}_i']$ are used as initial guess to exploit the optimal weights by minimizing the overall regression error.

6.3.2.2 *Deep feature space partitioning*

The extracted deep features are then used to classify working conditions via dividing the comprehensive state space. In this study, the division of comprehensive state space is achieved in a sequential manner. The deep feature space is first divided. Then, the division of deep feature space is mapped back to the division of the comprehensive state space. In this section, the division of deep feature space is introduced, which involves a rough division and a fine division. In the rough division, a k-dimensional tree (KD-Tree) is applied to first divide the space into different partitions [16][17]. In the fine division, the result of rough division is fed as initial guess of an LR classifier to fine-tune the classification [18].

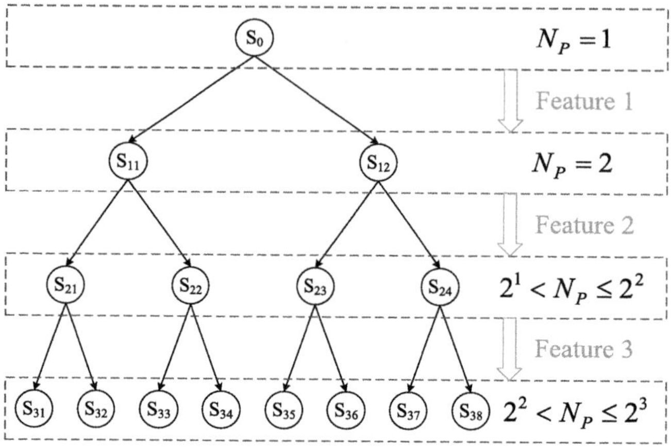

FIGURE 6.11 Rough division using a KD-Tree.

Rough division using a KD-Tree

KD-Tree is a variant of the binary tree. KD-Tree splits the deep feature space into two subspaces at each level iteratively until the deep feature space is divided into the desired number of subspaces (Fig. 6.11). The detailed steps of KD-Tree-based rough division include:

Step 1: Determine the desired numbers of partitions N_P.

Step 2: Calculate the number of required features and the level of splits. If $2^{N-1} < N_P \leqslant 2^N$, $N \geqslant 1$, then N features and N_S number of splits are required. We have

$$N_S = (2^{N-1} - 1) + (N_P - 2^{N-1}) = N_P - 1. \tag{6.47}$$

Step 3: Normalize the deep features. Calculate the variances of the normalized features, and rank the features according to their variances. Let $j = 1$.

Step 4: Calculate the median value of the jth feature. Consider the jth level, for partition $\mathrm{S}_{(j-1)q}$, where $q = 1$ to $2^{(j-1)}$, and generate a perpendicular splitting hyperplane which takes the median value on the axis of the jth feature. Then, the partition $\mathrm{S}_{(j-1)q}$ is divided into two new subspaces $\mathrm{S}_{j(2q-1)}$, $\mathrm{S}_{j(2q)}$. These two new subspaces are ranked according to the variance of the $(j+1)$th feature.

Step 5: If $j < N - 1$, then let $j = j + 1$, and repeat **Step 4** until $j = N - 1$.

Step 6: If $2^j < N_P$, then from $q = 1$ to $q = 2^j$, divide the partition S_{jq} using the Nth feature until $2q + (2^j - q) = N_P$.

After these steps, the original feature space S_0 is partitioned into $\{\mathrm{S}_{N1}, \mathrm{S}_{N2}, \cdots, \mathrm{S}_{N(2q)}, \mathrm{S}_{j(q+1)}, \cdots, \mathrm{S}_{j(2^j)}\}$ (denoted as $\{\mathrm{S}^{(1)}, \mathrm{S}^{(2)}, \cdots, \mathrm{S}^{(N_P)}\}$ hereafter).

Remark 1. In the rough division, the variance of deep features is selected as a division criterion. For each feature, larger feature variance corresponds to a wider distribution of feature values and therefore has a higher division efficiency [19].

Fine division based on LR

The KD-Tree approach roughly divides the deep feature space into different partitions, which can be used as an initial guess for the fine division. The fine division is achieved using LR classifier. It estimates and discriminates among the probabilities that the current process belongs to different working conditions:

$$P(y|\mathbf{z}) = \frac{1}{1+\mathrm{e}^{-\sigma(\mathbf{z})}}, \tag{6.48}$$

$$\sigma(\mathbf{z}) = \beta_0 + \sum_{i=1}^{N} \beta_i z_i, \tag{6.49}$$

where $P(y|\mathbf{z})$ is the probability that the current observation $\mathbf{z}$ belongs to the working condition with label y. If $P(y|\mathbf{z}) > 0.5$, i.e., $\sigma(\mathbf{z}) > 0$, then $\mathbf{z}$ is considered as belonging to working condition y; $\sigma(\mathbf{z}) = 0$ is the decision boundary; $\boldsymbol{\beta} = [\beta_0, \beta_1, \cdots, \beta_N]$ is the coefficient set to be identified for each working condition.

The steps of fine division include:

Step 1: Let $r = 1$.

Step 2: For subspace $S^{(r)}$, identify the coefficient set $\boldsymbol{\beta}^{(r)}$ by minimizing the label estimation error:

$$\begin{aligned}\ell_{\mathrm{LR}}^{(r)}(\boldsymbol{\beta}^{(r)}) &= \sum_{j=1}^{M}[\frac{1}{1+\mathrm{e}^{-\sigma^{(r)}(\mathbf{z}_j)}} - y^{(j)}]^2 \\ &= \sum_{j=1}^{M}[\frac{1}{1+\mathrm{e}^{-[\beta_0^{(r)}+\sum_{i=1}^{N}\beta_i^{(r)}z_i^{(j)}]}} - y^{(j)}]^2,\end{aligned} \tag{6.50}$$

where $y^{(j)}$ is the label of the jth sample given by rough division. The subspace surrounded by surface $\sigma^{(r)}(\mathbf{z}) = 0$ is the rth subspace of the entire feature space.

Step 3: If $r < N_P$, $r = r + 1$, repeat **Step 2.**

Remark 2. The partition of the deep feature space can be mapped back to the partition of the comprehensive state space. However, the borders of the subspaces are not time-invariant. In addition, new types of working conditions may emerge during the production process. If a sufficient amount of data samples associated with new working conditions is generated, the deep feature space division should be repeated to update the partition.

6.3.3 Model performance evaluation

In order to test the model performance, data samples collected from the site were utilized to conduct an experimental study. A number of 1050 data samples were used in the experimental study. For confidential reasons, the data are scaled and desensitized. Among the data samples, 1000 samples were used for working condition classification, training the machine learning submodels, and parameter identification of the first-principle submodels. The rest are used in model performance evaluation.

The aim of the study is to test the overall performance of the integrated model framework. The steps of the evaluation include:

Step 1: Deep feature extraction.
Step 2: Deep feature space partitioning.
Step 3: Identifying the model parameters of the FPM (nominal kinetic model) and training the machine learning compensation model under different working conditions.
Step 4: Comparing the performance of the nominal kinetic model, the pure machine learning-based model, and the integrated modeling approach.

The integrated modeling framework is based on the CSS descriptive system. Therefore, the inlet conditions, reaction conditions, and output states are selected according to the reaction mechanism.

(i) Output states: The cobalt ion outlet concentration of each reactor.
(ii) Inlet conditions: Flow rate, pH, and concentrations of the metallic ions (e.g., Co^{2+}, Cu^{2+}, Zn^{2+}, etc.) in the feed solution.
(iii) Reaction conditions: ORP of each reactor, flow rate of the arsenic trioxide and spent acid.

To sum up, there are 4 output states, 12 variables indicating inlet conditions, and 6 variables for the reaction conditions (Table 6.2). Therefore, the total dimension of output states, inlet conditions, and reaction conditions is 22. The correlations between inlet conditions, reaction conditions, and output states are shown in Table 6.3.

TABLE 6.2 Variables selected for the CSS.

State type	Variables
Output states	c_1, c_2, c_3, c_4
Reaction conditions	$v_1, v_2, v_3, v_4, f_{As}, f_{Acid}$
Inlet conditions	$f_0, m_{pH}, c_0, c_{Cu}, c_{Zn}, c_{Cd}, c_{Ni}, c_{As}, c_{Sb}, c_{Ge}, c_{Fe}, c_{Ca}$

In the working condition classification, the number of working conditions is first selected as $N_P = 7$. Therefore, the required number of deep features for classification is three. Two levels of auto-encoders are stacked to form an SAE for deep feature extraction. The number of latent variables in the two auto-encoders is 18 and 3, respectively. The activation function adopted is the

TABLE 6.3 The correlations between the condition variables and output states.

Variable	c_1	c_2	c_3	c_4
v_1	0.2143	0.1476	0.1572	0.3085
v_2	0.0580	0.0419	0.0206	−0.0863
v_3	−0.1221	−0.1567	−0.2421	−0.1648
v_4	0.1511	0.0595	0.0494	0.0129
f_{As}	0.3712	0.3427	0.3759	0.3942
f_{Acid}	0.0128	−0.0555	0.0420	0.1625
f_O	0.3467	0.2015	0.2529	0.2664
m_{pH}	0.1390	0.0683	0.0469	0.0354
c_O	0.5910	0.3890	0.3085	0.4931
c_{Cu}	0.1819	0.2220	0.2317	0.3073
c_{Zn}	0.0481	0.0938	0.0696	0.0772
c_{Cd}	−0.2684	−0.1675	−0.1493	−0.1023
c_{Ni}	0.0182	−0.0515	−0.0377	−0.0005
c_{As}	0.1889	0.1144	0.1432	0.2587
c_{Sb}	−0.0666	−0.0958	−0.0936	−0.0247
c_{Ge}	−0.1512	−0.2045	−0.1987	−0.1254
c_{Fe}	0.1264	0.0946	0.0443	−0.0388
c_{Ca}	0.1299	0.0040	−0.0385	−0.0737

sigmoid function:

$$f_{sig}(x) = \frac{1}{1 + e^{-x}}. \tag{6.51}$$

After the deep feature extraction, rough division was conducted. The subspaces were denoted as S_{31} (working condition type 1), S_{32} (type 2), S_{33} (type 3), S_{34} (type 4), S_{35} (type 5), S_{36} (type 6), and S_{24} (type 7) (Figs. 6.11 and 6.12). Then, fine division was conducted. In fine division, LR provides the probabilities that the current working condition belongs to each working condition. The probability information is useful especially for the junctions of two or more subspaces where the type of working condition is hard to determine (Fig. 6.12). The deep feature space division result is shown in Fig. 6.12. The locations of the 50 test samples in the deep feature space are also denoted. It can be observed that in the middle of the subspace, the working condition type is certain, and the probability it belongs to the working condition associated with the subspace is 100%. However, at the border of different subspaces, a working point may exhibit the dynamic features of different working conditions. For example, the 30th sample is located on the junction of subspaces S35 and S36. The probabilities that it belongs to working conditions type 5 and type 6 are 85% and 15%, respectively. The corresponding partitioning of the comprehensive state space is shown in Fig. 6.13.

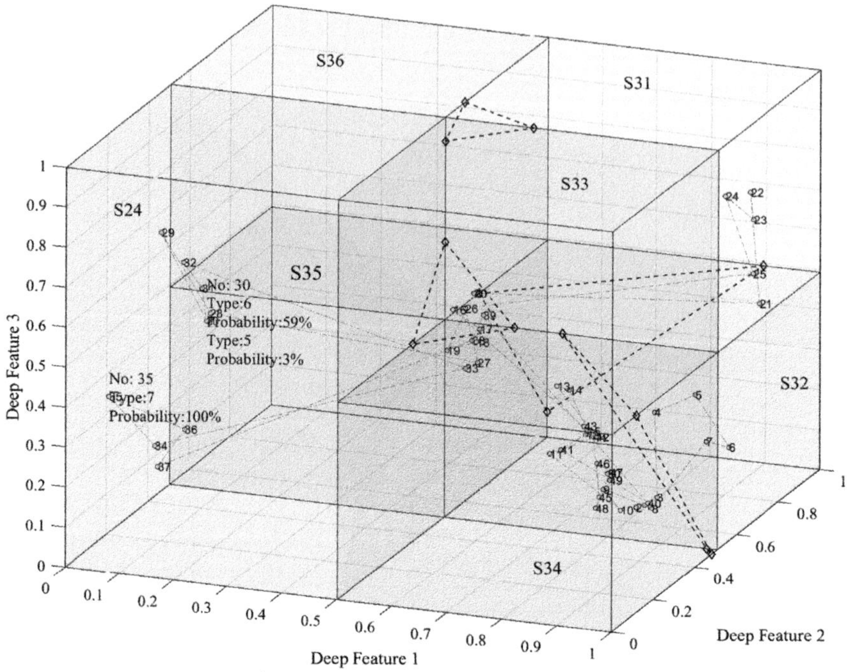

FIGURE 6.12 Division of the deep feature space.

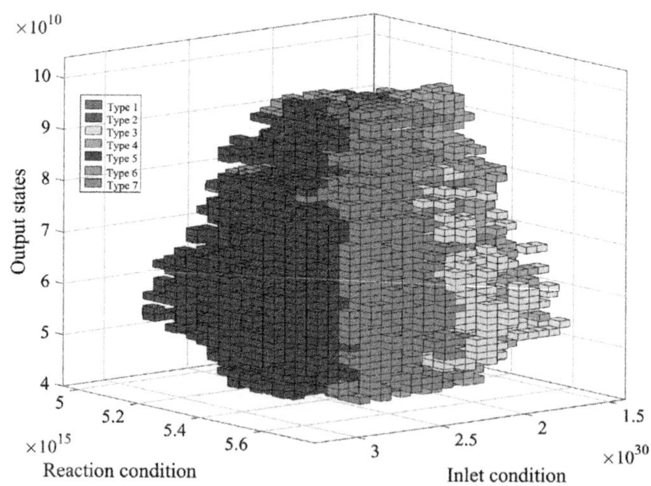

FIGURE 6.13 Division of the comprehensive state space.

For each reactor, the nominal kinetic model part of Eq. (6.42) can be derived using Eq. (6.41). The parameters of the kinetic model under each working condition are identified using the data samples associated with the corresponding

TABLE 6.4 Parameters of the nominal kinetic model under different working conditions.

Working condition	Reactor	A_β	E_e	γ	v_{eq}
S31	Reactor 1	6981106	72932	0.618	−0.300
	Reactor 2	7286393	79847	0.625	−0.332
	Reactor 3	8073669	79998	0.706	−0.373
	Reactor 4	8983144	71945	0.664	−0.436
S32	Reactor 1	7362697	72747	0.600	−0.300
	Reactor 2	7832393	79968	0.689	−0.367
	Reactor 3	8668537	77202	0.671	−0.392
	Reactor 4	6562077	72362	0.603	−0.404
S33	Reactor 1	5984033	72400	0.600	−0.300
	Reactor 2	8365405	78822	0.601	−0.348
	Reactor 3	5308410	79043	0.608	−0.348
	Reactor 4	8186106	72876	0.750	−0.463
S34	Reactor 1	7638676	73391	0.600	−0.300
	Reactor 2	8858660	77157	0.797	−0.427
	Reactor 3	6213033	77216	0.791	−0.429
	Reactor 4	5874675	78829	0.621	−0.368
S35	Reactor 1	5957729	72265	0.600	−0.300
	Reactor 2	7970208	79557	0.630	−0.352
	Reactor 3	8510620	77102	0.734	−0.417
	Reactor 4	8817141	73734	0.695	−0.440
S36	Reactor 1	5349874	72240	0.600	−0.300
	Reactor 2	6516429	74700	0.616	−0.385
	Reactor 3	5255406	78698	0.772	−0.414
	Reactor 4	7004144	71521	0.704	−0.457
S24	Reactor 1	5311958	72624	0.600	−0.300
	Reactor 2	5438328	77854	0.671	−0.382
	Reactor 3	8817776	70562	0.617	−0.440
	Reactor 4	8756903	79996	0.640	−0.379

working condition and shown in Table 6.4. The prediction result using the nominal kinetic model is presented in Fig. 6.14 and Table 6.5. For comparison, the outlet cobalt ion concentrations of each reactor are also predicted using a pure data-driven model (PDDM). In the PDDM, the deep features are first extracted using an SAE. Then, the deep features are fed to RBFNN which outputs the prediction of outlet cobalt ion concentrations of each reactor. It can be observed from the results that compared with the kinetic model, PDDM is less sensitive to the change of working conditions.

To construct the integrated model, for a reactor, a deviation term $c^{\text{deviation}}$ is added to the nominal kinetic model. We have

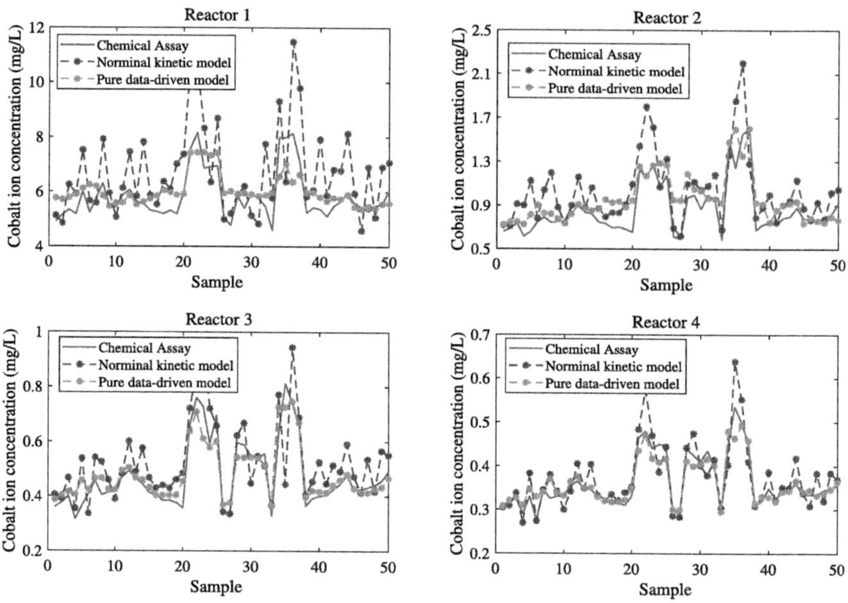

FIGURE 6.14 Prediction results of the nominal kinetic model and the pure data-driven model.

$$c(t) = c(0) + \int_0^t \left(\frac{f_{\text{in}}(\tau)}{V} c_{\text{in}} - \frac{f_{\text{out}}}{V} c(\tau) - A_\beta(i)\exp\left(-\frac{E_e(i) + 2\alpha(i)\mathrm{F}(e_{\text{orp}}(\tau) - e_{\text{eq}}(i))}{\mathrm{R}T_c}\right)c(\tau) \right) d\tau + c^{\text{deviation}}. \tag{6.52}$$

Similarly to the PDDM, the deviation term is modeled as a combination of a neural network and an SAE. The input of the neural network is the deep features extracted using SAE. Its output is the estimation of the deviation between the output of a nominal kinetic model and the practical value of the outlet cobalt ion concentration. The prediction result of the integrated model is presented in Fig. 6.15. The comparison of results from the integrated model and the nominal kinetic model is presented in Table 6.5 and Fig. 6.16 [15]. It can be observed from the comparison that the integrated modeling framework has a higher average accuracy and more stable prediction performance. The improvement in modeling accuracy is brought about by the higher information utilization rate of the integrated modeling framework. By combining reaction kinetics, production data, and information about the current working condition, the system dynamics can be described more comprehensively. In addition, as the integrated modeling approach is based on the CSS framework, other properties, e.g., controller gain, status, and suggested operation, can also be associated with the current working point to provide a comprehensive description of the process (Fig. 6.17).

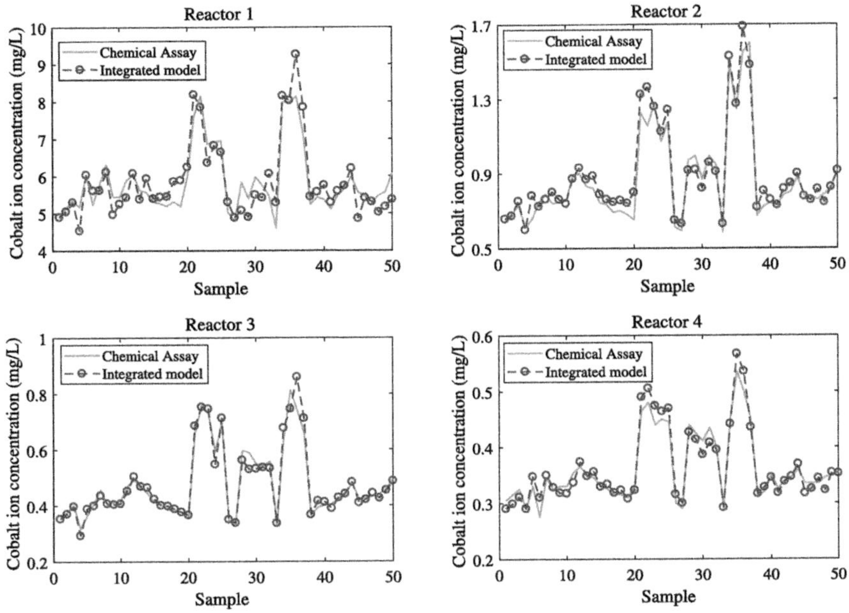

FIGURE 6.15 Prediction results of the integrated modeling framework.

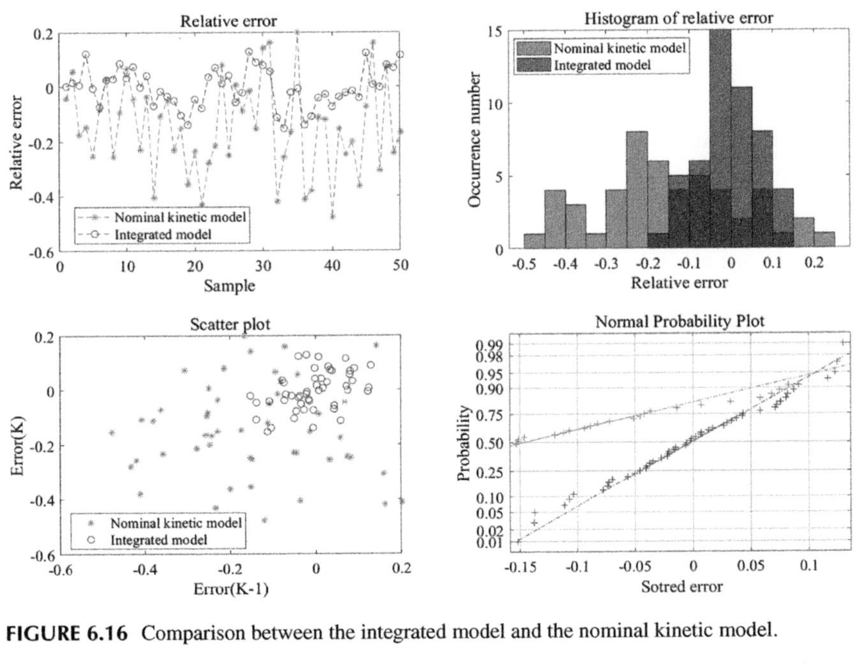

FIGURE 6.16 Comparison between the integrated model and the nominal kinetic model.

TABLE 6.5 Performance comparison between the nominal kinetic model (FPM), the pure data-driven model (PDDM), and the integrated model using the average relative error (ARE) and the root mean square error (RMSE).

Performance measure	Reactor 1	Reactor 2	Reactor 3	Reactor 4
ARE of FPM	15.19%	15.55%	16.14%	17.21%
ARE of PDDM	8.38%	10.70%	6.67%	8.32%
ARE of integrated model	5.7%	4.02%	3.33%	7.90%
RMSE of FPM	1.3967	0.6132	0.2506	0.0706
RMSE of PDDM	0.6296	0.3877	0.1069	0.0397
RMSE of integrated model	0.4166	0.1568	0.0679	0.0314

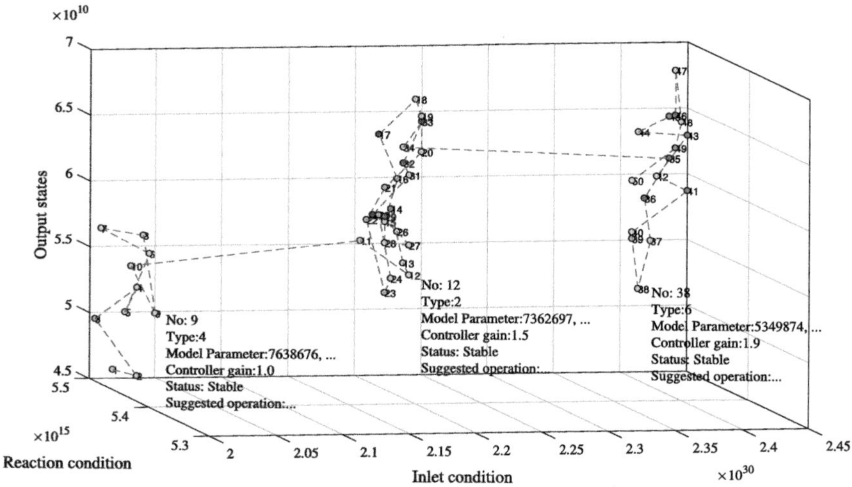

FIGURE 6.17 Evolution of ACP in CSS, as illustrated using test samples.

References

[1] V. Van der Pas, D.B. Dreisinger, A fundamental study of cobalt cementation by zinc dust in the presence of copper and antimony additives, Hydrometallurgy 43 (1) (1996) 187–205.

[2] M. Sadegh Safarzadeh, M.S. Bafghi, D. Moradkhani, M. Ojaghi Ilkhchi, A review on hydrometallurgical extraction and recovery of cadmium from various resources, Minerals Engineering 20 (3) (2007) 211–220.

[3] B. Sun, W. Gui, T. Wu, Y. Wang, C. Yang, An integrated prediction model of cobalt ion concentration based on oxidation reduction potential, Hydrometallurgy 140 (2013) 102–110.

[4] A. Dib, L. Makhloufi, Mass transfer correlation of simultaneous removal by cementation of nickel and cobalt from sulphate industrial solution containing copper: part II: onto zinc powder, Chemical Engineering Journal 123 (2006) 53–58.

[5] S. Fugleberg, A. Jarvinen, E. Yllo, Recent development in solution purification at Outokumpu zinc plant, Kokkola, World Zinc 93 (1) (1993) 241–247.

[6] D. Yang, G. Xie, G. Zeng, J. Wang, R. Li, Mechanism of cobalt removal from zinc sulfate solutions in the presence of cadmium, Hydrometallurgy 81 (1) (2006) 62–66.

[7] S.G. Fountoulakis, Studies on the cementation of cobalt with zinc in the presence of copper and antimony additives, Ph.D. thesis, Columbia University, New York, US, 1983.

[8] A.M. Polcaro, S. Palmas, S. Dernini, Kinetics of cobalt cementation on zinc powder, Industrial & Engineering Chemistry Research 34 (9) (1995) 3090–3095.
[9] L.I. Antropov, Theoretical Electrochemistry, Imported Pubn, 1972.
[10] J.C. Balarini, L. de Oliveira Polli, T.L. Santos Miranda, R.M.Z. de Castro, A. Salum, Importance of roasted sulphide concentrates characterization in the hydrometallurgical extraction of zinc, Minerals Engineering 21 (1) (2008) 100–110.
[11] B. Sun, C. Yang, Y. Wang, W. Gui, C. Ian, O. Laurentz, A comprehensive hybrid first principles/machine learning modeling framework for complex industrial processes, Journal of Process Control 86 (2020) 30–43.
[12] Y. LeCun, Y. Bengio, G. Hinton, Deep learning, Nature 521 (7553) (2015) 436–444.
[13] Y. Chen, H. Jiang, C. Li, X. Jia, P. Ghamisi, Deep feature extraction and classification of hyperspectral images based on convolutional neural networks, IEEE Transactions on Geoscience and Remote Sensing 54 (10) (2016) 6232–6251.
[14] C. Shang, F. Yang, D. Huang, W. Lyu, Data-driven soft sensor development based on deep learning technique, Journal of Process Control 24 (3) (2014) 223–233.
[15] X. Yuan, B. Huang, Y. Wang, C. Yang, W. Gui, Deep learning based feature representation and its application for soft sensor modeling with variable-wise weighted SAE, IEEE Transactions on Industrial Informatics 14 (2018) 3235–3243.
[16] J.L. Bentley, Multidimensional binary search trees used for associative searching, Communications of the ACM 18 (9) (1975) 509–517.
[17] J. Zhang, H. Guo, F. Hong, X. Yuan, T. Peterka, Dynamic load balancing based on constrained K-D tree decomposition for parallel particle tracing, IEEE Transactions on Visualization and Computer Graphics 24 (1) (2018) 954–963.
[18] F.E. Harrell, Regression Modeling Strategies, Springer-Verlag, Cham, 2015.
[19] L. Lennart, System Identification: Theory for the User, second ed., Prentice Hall, Upper Saddle River, NJ, 1999.

Chapter 7

Intelligent optimal setting control of the cobalt removal process

Contents

7.1 Problem analysis

In this chapter, the optimization and control problem of ACP is discussed. For an nonferrous metallurgical process, its control strategy is generally designed according to the control objective, the configuration of controlled/manipulated variables, and the characteristics of the process.

In ACP, zinc powder and arsenic trioxide are added to reduce cobalt ions under specific reaction conditions [1]. The outlet cobalt ion concentration of the last reactor is the technical index, which reflects the cobalt removal performance. The total consumption of valuable zinc powder, which is related to production cost, is the economical index. The operation objective of ACP is threefold:

(i) The outlet cobalt ion concentration of the last reactor is lower than a predefined technical threshold.
(ii) Avoid large fluctuations of the process.
(iii) Decrease the zinc powder consumption.

In practice, the operators of ACP evaluate the running status and make operation decisions according to the value and trend of important process variables, e.g., flow rates of feeding zinc sulfate solution, recycled underflow, spent acid and arsenic trioxide, as well as ORP, zinc powder dosage, and temperature of each reactor. Among these process variables:

Modeling, Optimization, and Control of Zinc Hydrometallurgical Purification Process
https://doi.org/10.1016/B978-0-12-819592-5.00018-1

(i) The flow rate of arsenic trioxide is determined according to the chemometric relation discussed in Section 6.2.1.
(ii) The flow rates of spent acid and recycled underflow are determined in the trial operation stage, and kept constant in most cases during production.
(iii) The flow rate of feeding zinc sulfate solution is not only determined by ACP. Instead, it is mainly determined by the requirement on zinc metal yield, and adjusted according to the running status of the overall zinc hydrometallurgy process.

Therefore, the zinc powder dosages of each reactor are the most frequently adjusted variables [2].

ACP is a multiphase reaction process with various influencing factors. The relationship among technical index, manipulated variables, and monitored process variables exhibits complex characteristics, such as nonlinearity and strong coupling. In addition, with the shrinking of high-grade minerals and supply-side volatility, the zinc smelting plants purchase zinc concentrates from different mines. Therefore, the composition and physicochemical properties of zinc concentrate are complex and time-varying. As a result, the working conditions of ACP fluctuate. By routine operation, human operators gained rich experience which can guide their operation. So their operation is reasonable in most cases. However, the optimality of their operation cannot be guaranteed.

Generally, under different working conditions, the operation objective of ACP is different. Therefore, different optimization and control strategies should be adopted. When ACP is under the normal operating state, i.e., the process is under steady-state conditions or there are no large fluctuations, economical optimization is the main objective. A strategy that minimizes the total zinc powder consumption of ACP is required. When ACP is in an abnormal operating state, i.e., the process is under unsteady-state conditions or large fluctuations emerge, the main objective is to recover the process to the normal operating state. The regulation of the process state plays a more important role compared with economical optimization.

Based on the above analysis, an optimization and control strategy based on the idea of "normal-state optimization, abnormal-state adjustment" is proposed [3]. The strategy accords with the "multiple working condition" feature of ACP. As shown in Fig. 7.1, the operating state of ACP is monitored online. For the normal-state optimization problem, the concepts of zinc powder utilization rate and cobalt removal ratio (CRR) are proposed. Based on these two concepts, a two-layer gradient optimization framework is established. Under normal-state conditions, the economical operation of ACP can be realized by optimizing the CRR according to the zinc powder utilization efficiency factor (ZPUF) of each reactor. The upper layer works at a lower frequency, optimizing the CRR of each reactor according to their zinc powder utilization rates, i.e., optimal setting of the decline gradient of the outlet cobalt ion concentration along the reactors. The lower layer works at a higher frequency, transferring the optimal setting value of the technical index to the value of manipulated variables, i.e., determination

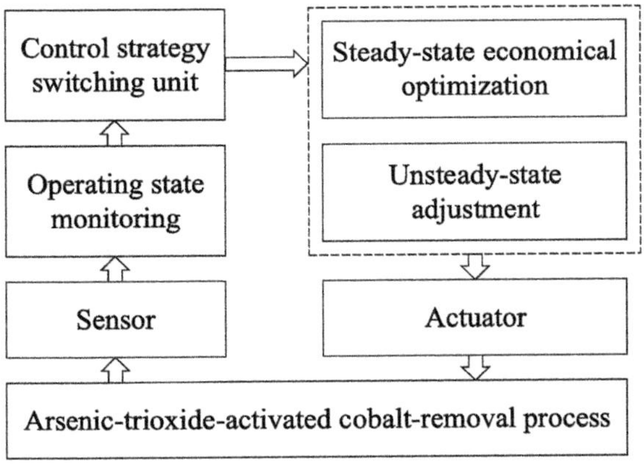

FIGURE 7.1 Intelligent optimal setting control strategy.

of the zinc powder dosage of each reactor. For the abnormal-state adjustment problem, rational setting values of control variables [4] are given by a case-based reasoning (CBR) controller [5], in which an appropriate solution for the new problem can be obtained through retrieving past successful experiences.

7.2 Normal-state economical optimization

7.2.1 Problem formulation

In normal-state conditions, the main concern is to minimize the production cost. The overall zinc powder dosage directly affects the economic and technical indices of ACP. An excessive zinc powder dosage is a waste of valuable resources, while an insufficient zinc powder dosage cannot achieve the desired cobalt removal performance [2]. According to the principle of material balance, the amount of zinc powder added to a reactor should be determined according to its efficiency in removing cobalt ions.

Assuming that the fluid in each reactor is completely mixed, the contents in the reactor are uniform. Then, for a single reactor, the mass balance of cobalt ions is

$$V\frac{dc}{dt} = Fc_{\text{in}} - Fc - Vr_c, \tag{7.1}$$

where c_{in} and c represent inlet and outlet cobalt ion concentrations, respectively, F means flux of inlet/outlet solutions, V is the volume of the reactor, and r_c is the removal rate of cobalt ions. If the reactor is under steady-state conditions, then

$$\frac{dc}{dt} = 0. \tag{7.2}$$

If $\frac{dc}{dt}$ is small enough, then the process is also considered as in the normal state. Combining Eq. (7.1) and Eq. (7.2),

$$r_c = \frac{F(c_{\text{in}} - c)}{V}. \tag{7.3}$$

Considering the reaction

$$As^{3+} + Co^{2+} + 2.5Zn \rightarrow CoAs + 2.5Zn^{2+},$$

under ideal conditions, i.e., 2.5 mol zinc powder can remove 1 mol cobalt ions, the removal rate of cobalt ions is the same as the reaction rate of zinc powder. So for a reactor with volume V, if the cobalt ion concentration is reduced from $c_{\text{in}}^{\text{Co}}$ to c^{Co}, then the zinc powder dosage required is

$$u_{\text{Ideal}}^{\text{Co}} = \frac{2.5M_{\text{Zn}}}{M_{\text{Co}}} r_c V t_{\text{retention}} = \frac{2.5M_{\text{Zn}}}{M_{\text{Co}}} V(c_{\text{in}}^{\text{Co}} - c^{\text{Co}}), \tag{7.4}$$

where $u_{\text{Ideal}}^{\text{Co}}$ is the weight of zinc powder that reacts with cobalt ions, M_{Zn} and M_{Co} are the atomic weights of Zn and Co, respectively, and $t_{\text{retention}} = \frac{V}{F}$ is the average retention time.

In practical ACP, zinc powder is not only involved in the replacement of cobalt ions, but also in the following reactions:

$$\begin{aligned}
Cu^{2+} + Zn &\rightarrow Zn^{2+} + Cu, \\
As^{3+} + 3Cu^{2+} + 4.5Zn &\rightarrow Cu_3As + 4.5Zn^{2+}, \\
As^{3+} + Co^{2+} + 2.5Zn &\rightarrow CoAs + 2.5Zn^{2+}, \\
As^{3+} + Ni^{2+} + 2.5Zn &\rightarrow NiAs + 2.5Zn^{2+}, \\
2H^{+} + Zn &\rightarrow Zn^{2+} + H_2 \uparrow.
\end{aligned}$$

Therefore, according to Eq. (7.4), considering a single reactor, its theoretical zinc powder dosage under steady-state conditions can be expressed by

$$u_{\text{Ideal}} = u_{\text{Ideal}}^{\text{Cu}} + u_{\text{Ideal}}^{\text{Co}} + u_{\text{Ideal}}^{\text{Ni}} + u_{\text{Ideal}}^{\text{H}}, \tag{7.5}$$

where $u_{\text{Ideal}}^{\text{Cu}}$, $u_{\text{Ideal}}^{\text{Ni}}$, and $u_{\text{Ideal}}^{\text{H}}$ denote zinc powder consumed in reaction with Cu^{2+}, Ni^{2+}, and H^{+}:

$$u_{\text{Ideal}}^{\text{Cu}} = \frac{(1.5 - 0.5\alpha)M_{\text{Zn}}}{M_{\text{Cu}}} V(c_{\text{in}}^{\text{Cu}} - c^{\text{Cu}}), \tag{7.6}$$

$$u_{\text{Ideal}}^{\text{Ni}} = \frac{2.5M_{\text{Zn}}}{M_{\text{Ni}}} V(c_{\text{in}}^{\text{Ni}} - c^{\text{Ni}}), \tag{7.7}$$

$$u_{\text{Ideal}}^{\text{H}} = \frac{0.5M_{\text{Zn}}}{M_{\text{H}}} V(c_{\text{in}}^{\text{H}} - c^{\text{H}}), \tag{7.8}$$

where M_{Cu}, M_{Ni}, and M_{H} are the atomic weight of Cu, Ni, and H; $c_{\mathrm{in}}^{\mathrm{Cu}}$, $c_{\mathrm{in}}^{\mathrm{Ni}}$, and $c_{\mathrm{in}}^{\mathrm{H}}$ and c^{Cu}, c^{Ni}, and c^{H} are the inlet and outlet concentrations of Cu^{2+}, Ni^{2+}, and H^{+}, respectively; α is the proportion of copper ions taking part in the reaction $Cu^{2+} + Zn \rightarrow Zn^{2+} + Cu$ $(0 \le \alpha \le 100\%)$.

However, the reaction between cobalt ions and zinc powder is a stochastic process with kinetic barriers [1]. The reaction rate is influenced by many factors, such as pH and viscosity. When the local pH of the solution is high, it is easy to generate basic zinc sulfate. The basic zinc sulfate sticks on the surface of zinc powder particles, which restricts the contact between zinc powder particle and cobalt ions. In addition, when the concentration of zinc ions in the solution is high, it is difficult for elemental zinc in the zinc powder to lose electrons and diffuse into the solution as zinc ions. Therefore, zinc powder is not fully involved in the cobalt removal reactions, i.e., the utilization efficiency of zinc powder is not 100%.

Consider an ACP composed of N reactors. If we denote $\mu_{\mathrm{Co}}(i)$ as the ZPUF in reaction with cobalt ions of the ith reactor, then the practical zinc powder dosage in replacing all the cobalt ions is

$$\begin{aligned} u_{\mathrm{Real}}^{\mathrm{Co}} &= \frac{2.5M_{\mathrm{Zn}}}{M_{\mathrm{Co}}} V \sum_{i=1}^{N} \mu_{\mathrm{Co}}^{-1}(i)[c^{\mathrm{Co}}(i) - c^{\mathrm{Co}}(i+1)] \\ &= \frac{2.5M_{\mathrm{Zn}}}{M_{\mathrm{Co}}} V c_{\mathrm{in}}^{\mathrm{Co}} \sum_{i=1}^{N} \mu_{\mathrm{Co}}^{-1}(i)\lambda_{\mathrm{Co}}(i) \\ &= U_{\mathrm{Co}} \sum_{i=1}^{N} \mu_{\mathrm{Co}}^{-1}(i)\lambda_{\mathrm{Co}}(i), \end{aligned} \tag{7.9}$$

where $c_{\mathrm{in}}^{\mathrm{Co}}$ is the inlet cobalt ion concentration of reactor 1, $U_{\mathrm{Co}} = \frac{2.5M_{\mathrm{Zn}}}{M_{\mathrm{Co}}} V c_{\mathrm{in}}^{\mathrm{Co}}$ is the zinc powder dosage needed to remove all the cobalt ions under ideal conditions, $\mu_{\mathrm{Co}}^{-1}(i)$ means the zinc powder utilization efficiency of reactor i $(i = 1, 2, 3, \cdots, N)$, and $\lambda_{\mathrm{Co}}(i)$ means the CRR of the ith reactor $(i = 1, 2, 3, \cdots, N)$. Therefore, similarly, considering all the reactions zinc powder is involved in, the overall zinc powder consumption is

$$\begin{aligned} u_{\mathrm{Real}} &= u_{\mathrm{Real}}^{\mathrm{Cu}} + u_{\mathrm{Real}}^{\mathrm{Co}} + u_{\mathrm{Real}}^{\mathrm{Ni}} + u_{\mathrm{Real}}^{\mathrm{H}} \\ &= U_{\mathrm{Co}} \sum_{i=1}^{N} \mu_{\mathrm{Co}}^{-1}(i)\lambda_{\mathrm{Co}}(i) + U_{\mathrm{Ni}} \sum_{i=1}^{N} \mu_{\mathrm{Ni}}^{-1}(i)\lambda_{\mathrm{Ni}}(i) \\ &\quad + U_{\mathrm{Cu}} \sum_{i=1}^{N} \mu_{\mathrm{Cu}}^{-1}(i)\lambda_{\mathrm{Cu}}(i) + U_{\mathrm{H}} \sum_{i=1}^{N} \mu_{\mathrm{H}}^{-1}(i)\lambda_{\mathrm{H}}(i), \end{aligned} \tag{7.10}$$

where $u_{\mathrm{Real}}^{\mathrm{Cu}}$, $u_{\mathrm{Real}}^{\mathrm{Ni}}$, and $u_{\mathrm{Real}}^{\mathrm{H}}$ are the overall real zinc powder consumption in replacing Cu^{2+}, Ni^{2+}, and H^{+}, respectively; $\mu_{\mathrm{Cu}}(i)$, $\mu_{\mathrm{Ni}}(i)$, and $\mu_{\mathrm{H}}(i)$ denote the utilization efficiency of zinc powder in reaction with Cu^{2+}, Ni^{2+}, and H^+ of the ith reactor, respectively; $\lambda_{Cu}(i)$, $\lambda_{Ni}(i)$, and $\lambda_H(i)$ are the consumption ratio of Cu^{2+}, Ni^{2+}, and H^+ in the ith reactor, respectively; and U_{Co}, U_{Ni}, and U_H are the zinc powder consumption in removing all the Cu^{2+}, Ni^{2+}, and H^+ under ideal conditions, respectively, with

$$U_{Cu} = \frac{(1.5 - 0.5\alpha)M_{Zn}}{M_{Cu}} V c_{in}^{Cu},$$
$$U_{Ni} = \frac{2.5 M_{Zn}}{M_{Ni}} V c_{in}^{Ni},$$
$$U_H = \frac{0.5 M_{Zn}}{M_H} V c_{in}^{H}.$$

The utilization efficiency of zinc powder in removing an impurity is related to the reaction rate. If the reaction rate is high, there will be more impurity ions removed per unit time and the zinc powder utilization efficiency is high. If the reaction rate is low, fewer impurity ions will be removed per unit time and the zinc powder utilization efficiency is low. Among Co^{2+}, Cu^{2+}, Ni^{2+}, and H^+, Cu^{2+} and H^+ are the easiest to react with zinc powder. Although the addition of arsenic trioxide made the precipitation of cobalt ions possible and the presence of copper ions accelerates the precipitation process, the activation energy in cobalt precipitation is still high. This indicates that the reaction rate of the cobalt precipitation reaction is still low. Therefore, zinc powder can replace far more copper and hydrogen ions than cobalt and nickel ions per unit time. Generally, the zinc powder utilization efficiencies in replacing copper ions and hydrogen ions are considered to be 100%. On the other hand, as nickel and cobalt are both iron group elements. They have very similar physical and electrochemical properties. Their atomic weights are 58.93 and 58.69, respectively. Their standard electrode potentials are −0.277 V and −0.230 V, respectively. They both exhibit the overvoltage phenomenon when precipitating on the surface of the zinc powder particle. The precipitation reactions of cobalt and nickel have high activation energy. So the zinc powder utilization efficiencies in replacing cobalt ions and nickel ions are considered as the same. This is also confirmed in the laboratory data. It is observed that the removal ratios of cobalt and nickel ions in each reactor are close. The ZPUF therefore concerns the zinc powder utilization in replacing cobalt and nickel ions.

7.2.1.1 Zinc powder utilization efficiency factor

Zinc powder utilization efficiency factor: For the ith reactor ($i = 1, 2, \cdots, N$), consider the retention interval of a block of solution with volume V. The zinc powder utilization efficiency factor [2] is the ratio between the theoretical zinc powder consumption in replacing Co^{2+} and Ni^{2+} and the real zinc powder consumption:

$$\mu_{\mathrm{Co}}(i)=\frac{u_{\mathrm{Ideal}}^{\mathrm{Co}}(i)+u_{\mathrm{Ideal}}^{\mathrm{Ni}}(i)}{u_{\mathrm{Real}}(i)-u_{\mathrm{Real}}^{\mathrm{Cu}}(i)-u_{\mathrm{Real}}^{\mathrm{H}}(i)}=\frac{u_{\mathrm{Ideal}}^{\mathrm{Co}}(i)+u_{\mathrm{Ideal}}^{\mathrm{Ni}}(i)}{u_{\mathrm{Real}}^{\mathrm{Co}}(i)+u_{\mathrm{Real}}^{\mathrm{Ni}}(i)}, \tag{7.11}$$

where $u_{\mathrm{Ideal}}^{\mathrm{Co}}(i)$ and $u_{\mathrm{Ideal}}^{\mathrm{Ni}}(i)$ are the zinc powder consumption in replacing Co^{2+} and Ni^{2+} of the ith reactor, $u_{\mathrm{Real}}^{\mathrm{Co}}(i)$ and $u_{\mathrm{Real}}^{\mathrm{Ni}}(i)$ are the real zinc powder consumption in replacing Co^{2+} and Ni^{2+} of the ith reactor, $u_{\mathrm{Real}}(i)$ is the overall zinc powder consumption in the ith reactor, and $u_{\mathrm{Real}}^{\mathrm{Cu}}(i)$ and $u_{\mathrm{Real}}^{\mathrm{H}}(i)$ are the real zinc powder consumption in replacing Co^{2+} and Ni^{2+} of the ith reactor.

According to Eq. (7.11), the overall zinc powder consumption in the ith reactor can be expressed as

$$u_{\mathrm{Real}}(i)=u_{\mathrm{Real}}^{\mathrm{Cu}}(i)+u_{\mathrm{Real}}^{\mathrm{H}}(i)+\mu_{\mathrm{Co}}^{-1}[u_{\mathrm{Ideal}}^{\mathrm{Co}}(i)+u_{\mathrm{Ideal}}^{\mathrm{Ni}}(i)]. \tag{7.12}$$

As cobalt and nickel have similar electrochemical properties, i.e., close standard electrode potentials, reaction rates, and atomic weights [6], if $c^{\mathrm{Co}}(i)=k_r c^{\mathrm{Ni}}(i)$, then Eq. (7.12) can be approximated as

$$u_{\mathrm{Real}}(i)=u_{\mathrm{Real}}^{\mathrm{Cu}}(i)+u_{\mathrm{Real}}^{\mathrm{H}}(i)+(1+k_r)\mu_{\mathrm{Co}}^{-1}(i)u_{\mathrm{Ideal}}^{\mathrm{Co}}(i). \tag{7.13}$$

Considering the retention interval of the solution in all the reactors of ACP, if the ACP is under normal-state conditions, then the overall zinc powder consumption can be formulated as

$$\begin{aligned}u_{\mathrm{Real}}&=u_{\mathrm{Real}}^{\mathrm{Cu}}+u_{\mathrm{Real}}^{\mathrm{H}}+(1+k_r)\sum_{i=1}^{N}\mu_{\mathrm{Co}}^{-1}(i)u_{\mathrm{Ideal}}^{\mathrm{Co}}(i)\\&=u_{\mathrm{Real}}^{\mathrm{Cu}}+u_{\mathrm{Real}}^{\mathrm{H}}+(1+k_r)\frac{2.5M_{\mathrm{Zn}}}{M_{\mathrm{Cu}}}V\sum_{i=1}^{N}\mu_{\mathrm{Co}}^{-1}(i)[c^{\mathrm{Co}}(i)-c^{\mathrm{Co}}(i+1)].\end{aligned} \tag{7.14}$$

7.2.1.2 Cobalt removal ratio

Cobalt removal ratio: The CRR of a single reactor presents the ratio between the removed cobalt ions and the inlet cobalt ion concentration of the first reactor under normal-state conditions:

$$\lambda_{\mathrm{Co}}(i)=\frac{c^{\mathrm{Co}}(i)-c^{\mathrm{Co}}(i+1)}{c_{\mathrm{in}}^{\mathrm{Co}}}, \tag{7.15}$$

where $c^{\mathrm{Co}}(i)$ means inlet cobalt ion concentration of reactor i $(i=1,2,3,\cdots,N-1)$, and $c^{\mathrm{Co}}(1)=c_{\mathrm{in}}^{\mathrm{Co}}$, $c^{\mathrm{Co}}(N+1)$ is the outlet cobalt ion concentration of reactor N.

According to Eq. (7.15), Eq. (7.14) can be simplified as

$$u_{\mathrm{Real}} = u_{\mathrm{Real}}^{\mathrm{Cu}} + u_{\mathrm{Real}}^{\mathrm{H}} + (1+k_r)\frac{2.5M_{\mathrm{Zn}}}{M_{\mathrm{Cu}}} V c_{\mathrm{in}}^{\mathrm{Co}} \sum_{i=1}^{N} \mu_{\mathrm{Co}}^{-1}(i)\lambda_{\mathrm{Co}}(i). \tag{7.16}$$

If we denote

$$U'_{\mathrm{Co}} = (1+k_r)\frac{2.5M_{\mathrm{Zn}}}{M_{\mathrm{Cu}}} V c_{\mathrm{in}}^{\mathrm{Co}}, \tag{7.17}$$

then Eq. (7.16) can be further simplified as

$$u_{\mathrm{Real}} = u_{\mathrm{Real}}^{\mathrm{Cu}} + u_{\mathrm{Real}}^{\mathrm{H}} + U'_{\mathrm{Co}} \sum_{i=1}^{N} \mu_{\mathrm{Co}}^{-1}(i)\lambda_{\mathrm{Co}}(i). \tag{7.18}$$

The copper ion concentration in the feeding solution is determined before cobalt removal. The hydrogen ion concentration is determined by the flow rate and concentration of spent acid, which is usually kept constant. Therefore, if we only consider ACP, in Eq. (7.18), what can be optimized is the zinc powder consumption in replacing Co^{2+} and Cu^{2+}. If the cobalt and copper removal processes are considered together, then u_{Real} can be further optimized. In this chapter, we consider the former case.

7.2.1.3 Optimization problem formulation

Zinc powder utilization efficiency cannot only be used to estimate the required zinc powder dosage of a single reactor, but also to optimize the overall zinc powder consumption of ACP. In ACP, the cobalt ion concentration is gradually reduced along the reactors. Each reactor corresponds to a stage of the entire cobalt removal process and has different reaction states and conditions. Therefore, the zinc powder utilization efficiencies of the reactors are different. In addition, for a single reactor, due to fluctuations in the feeding conditions and the complex characteristics of the cobalt removal reaction, the reaction states and conditions inside each reactor are time-varying. Therefore, the zinc powder utilization efficiency of a single reactor is time-varying. If more cobalt is to be removed in the reactor with high zinc powder utilization efficiency and less cobalt is to be removed in the reactor with high zinc powder utilization efficiency during each optimization cycle, the overall zinc powder consumption of ACP can be optimized.

On the other hand, the CRR of a reactor is an external reflection of its internal reaction state. A higher CRR indicates that the reactor can remove more cobalt ions, and the rate of the cobalt removal reaction inside the reactor is higher. A lower CRR indicates that the reactor can remove fewer cobalt ions, and the rate of the cobalt removal reaction inside the reactor is low. In practice, each reactor has its own position and function in the cobalt removal process. For example, reactors #1 and #2 correspond to the rough adjustment stage of

ACP. These two reactors remove most of the cobalt ions. Reactors #3 and #4 correspond to the fine-tuning stage of the cobalt removal process. Only a small ratio of cobalt ions is removed in this stage. Moreover, the stability of the CRR reflects the stability of the reaction state inside the reactor. A small fluctuation of the CRR indicates that the reaction state inside the reactor is stable. A large fluctuation of the CRR indicates that the reaction state inside the reactor is unstable. Therefore, limiting the CRRs of each reactor in reasonable ranges not only conforms to the production requirements, but can also maintain the stable operation of ACP.

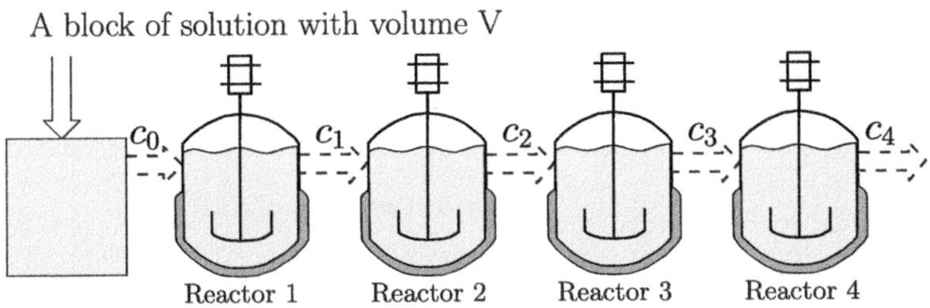

FIGURE 7.2 The retention of a block of solution with volume V in ACP.

Considering the retention interval of a block of solution with volume V from flowing into the first reactor to overflowing the last reactor (Fig. 7.2), the economical optimization problem of ACP under normal-state conditions can be transformed to a constraint optimization problem as follows:

$$\begin{aligned} \min \quad & u_{\text{Total}}(\lambda_1, \lambda_2, \cdots, \lambda_N) = U'_{\text{Co}} \sum_{i=1}^{N} \mu_{\text{Co}}^{-1}(i)\lambda_{\text{Co}}(i), \\ \text{st.} \quad & 0 \leqslant c_{\text{in}}^{\text{Co}}[1 - \sum_{i=1}^{N} \lambda_{\text{Co}}(i)] \leqslant c_{\text{index}}, \\ & \lambda_{\text{Co}}^{\text{low}}(i) \leqslant \lambda_{\text{Co}}(i) \leqslant \lambda_{\text{Co}}^{\text{up}}(i), \end{aligned} \tag{7.19}$$

in which c_{index} is the upper limit of the outlet cobalt ion concentration of the last reactor and $\lambda_{\text{Co}}^{\text{low}}(i)$ and $\lambda_{\text{Co}}^{\text{up}}(i)$ are the lower and upper limits of $\lambda_{\text{Co}}(i)$. The idea is to assign CRRs to the reactors according to their ZPUFs. The optimization objective is minimizing the overall zinc powder consumption. The constraints include:

(i) The CRRs of each reactor are within reasonable ranges, e.g., Table 7.1.
(ii) The outlet cobalt ion concentration of the last reactor is lower than the technical threshold.

TABLE 7.1 Reasonable range of CRR.

Reactor	Minimum CRR	Maximum CRR
1	50%	75%
2	25%	40%
3	1%	5%
4	1%	3%

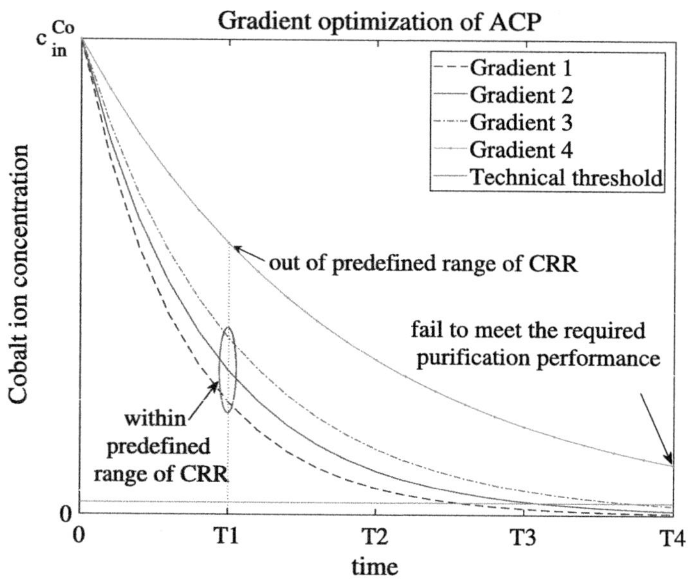

FIGURE 7.3 Gradient optimization of ACP.

Gradient optimization of ACP

In ACP, the cobalt ion concentration gradually decreases along the reactors, forming a curve with certain decline gradients (Fig. 7.3). The optimization of CRRs equals the optimization of the outlet cobalt ion concentration of each reactor. In this sense, the economical optimization problem of ACP equals optimizing the decline gradients of the outlet cobalt ion concentration along the reactors, i.e., finding the best decline curve from all possible decline curves.

Gradient optimization of ACP (GOACP): The economical optimization of ACP is essentially a gradient optimization problem, i.e., finding optimal decline gradients of cobalt ion concentration along the reactors according to the ZPUF of each reactor, which minimizes the overall zinc powder consumption of ACP and keeps the CRRs of each reactor within reasonable ranges and the outlet cobalt ion concentration of the last reactor lower than the technical threshold.

The decline gradient of cobalt ion concentration is very important to ACP. It affects not only the overall zinc powder consumption, but also the final cobalt ion concentration and stability of the operating state. As shown in Fig. 7.3, the green line indicates that if the cobalt removed in the first reactor is not sufficient, then the difficulty of cobalt removal in the subsequent reactors is increased, and hence the outlet cobalt ion concentration of the last reactor exceeds the technical threshold.

7.2.2 Two-layer gradient optimization under normal-state conditions

In order to realize the gradient optimization of ACP, a normal-state optimization framework as shown in Fig. 7.4 is established. The framework consists of two levels. The upper layer includes a CRR optimal setting unit which completes the optimization of the decline gradients of cobalt ion concentrations along the reactors. The lower layer includes a ZPUF estimation unit and a cobalt ion concentration tracking controller (ORP and zinc powder dosage controller). The working mechanism of the framework is as follows:

(i) Obtain the values of important process parameters of ACP, e.g., flow rates of inlet solution, arsenic trioxide and spent acid, the ORP and zinc powder dosage of each reactor, the BT value, concentrations of copper, cobalt, and nickel ions, and the outlet cobalt ion concentration of each reactor.

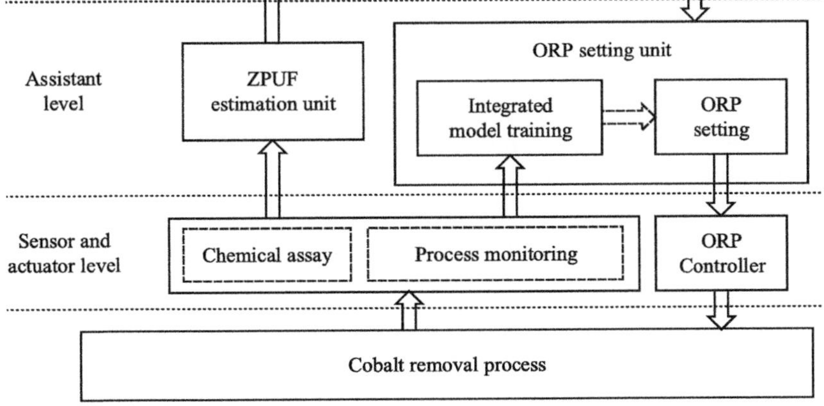

FIGURE 7.4 Normal-state optimal control frame.

(ii) The ZPUF estimation unit estimates the zinc powder utilization efficiency of each reactor according to the above important process variables.
(iii) The CRR optimal setting unit optimizes the CRR of each reactor according to Eq. (7.19) and obtains the optimal setting value of the outlet cobalt ion concentrations of each reactor.
(iv) The tracking controller of the cobalt ion concentration calculates the real-time ORP value or zinc powder dosage according to the optimal setting value of the outlet cobalt ion concentration of each reactor, which forces the outlet cobalt ion concentration of each reactor to stay around their setting values.

It can be seen from Fig. 7.4 that the ZPUF estimation unit, the CRR optimal setting unit, and the outlet cobalt ion concentration tracking controller are the core components of the gradient optimization framework. This framework optimizes the CRR of each reactor according to the inlet conditions and technical thresholds. It is aimed to minimize the overall zinc powder consumption and sustain the stability of the operation state. The cobalt removal rate of each reactor is optimally set to obtain the optimal gradient of cobalt ion concentration. It determines the value of ORP/zinc powder dosage according to the optimal setting values of outlet cobalt ion concentrations [7]. So compared with artificial operation, it is easier for the gradient optimization framework to find the optimal zinc powder dosage which minimizes the zinc powder consumption.

7.2.2.1 Online estimation of ZPUF

Due to the continuous fluctuations of the inlet cobalt and copper ion concentrations of reactor #1, the ACP exhibits different operating states. The zinc powder utilization efficiency is directly related to the operating state and is affected by various factors, e.g., zinc powder dosage, flow rates of spent acid and arsenic trioxide, the inlet concentrations of copper ions, etc. When the operating state is good, the cobalt removal rate is high and the zinc powder utilization efficiency is high. When the operating state is not good, the cobalt removal rate is low and the zinc powder utilization efficiency is low. Therefore, the zinc powder utilization efficiencies of each reactor are also time-varying in practice.

The ZPUF is defined regarding the zinc powder utilization performance on a time interval. More specifically, the average retention interval of the solution in the reactor is 1 to 2 hours. The real value of the ZPUF can be calculated only when the test results of impurity ion concentrations are obtained, which is time-consuming. Therefore, ZPUF cannot be calculated online. To overcome this disadvantage, a data-driven algorithm radical basis function neural network (RBFNN) is used to estimate the ZPUF [8]. In the offline stage, the regression model between the ZPUF and inlet/reaction conditions is trained. In the online stage, the real-time values of the inlet and reaction conditions are used as the input of the regression model to obtain the real-time ZPUF.

The RBFNN consists of three layers: input layer, hidden layer, and output layer. The inputs are process variables related to the technical index of concern. In the hidden layer of the RBFNN, a nonlinear activation function is used, i.e., a radial basis function. The output of the RBFNN is a weighted linear combination of the output of hidden neurons:

$$y = \sum_{i=1}^{m} \beta_i \psi_i(\chi), \tag{7.20}$$

where $\chi = [\chi_1, \chi_2, \cdots, \chi_n]^T$ and y are the input and output of the RBFNN, β_i $(i = 1, 2, \cdots, m)$ is the connecting weight between the ith neuron in the hidden layer and the output layer, and ψ_i is the output value of the ith neuron in the hidden layer. The most commonly used RBF is the Gaussian function:

$$\psi_i(\chi) = e^{-\|\chi - \mathbf{c}_i\|^2/\sigma_i^2}, \tag{7.21}$$

where $\|\chi - \mathbf{c}_i\|$ is the Euclidean distance between χ and $\mathbf{c}_i$, and $\mathbf{c}_i$ and σ_i are center and spread of the ith $(i = 1, 2, \cdots, m)$ node in the hidden layer.

The performance of an RBFNN is determined by the number of hidden neurons, the center and spread of each neuron, and the connecting weight $\beta = [\beta_1, \beta_2, \cdots, \beta_m]$. According to the way the centers are selected, there exist four types of training method, including random center selection, self-organized center selection, supervised center selection, and the orthogonal least squares (OLS) learning algorithm. In this chapter, the network is trained using the OLS method [9], while the aim is to minimize the following criterion:

$$\min \quad J = \sum_{i=1}^{P} |y_i - d_i|^2, \tag{7.22}$$

where P is the number of training data samples and d and y are the practical value and output of the RBFNN, respectively.

In order to test the performance of the RBFNN in estimating the ZPUF, a test experiment was conducted according to the following steps:

Step 1. Select the input variables and determine the number of hidden neurons and output: According to the study on the reaction mechanism, the input variables include:

- flow rate of the inlet solution;
- cobalt ion concentration of the inlet solution;
- copper ion concentration of the inlet solution;
- zinc powder dosage;
- flow rate of arsenic trioxide;
- ORP.

The number of neurons in the hidden layer is set as 15. The output is the ZPUF.

Step 2. Select the training data: The training data are selected from routinely collected production data of real plants. The training data should cover as many operating states as possible. However, the training data must be selected carefully, and data from the cases with an excessive additive dosage should be avoided.

Step 3. RBFNN configuration: Select the centers $\{\mathbf{c}_i\}$ using OLS, and choose the spread of each neuron as the closest Euclidean distance between its center and centers of other neurons. Adapt the connecting weights β using the adaptive gradient descent procedure [10].

Step 4. Performance test: Randomly choose 50 test data samples and compare the model output with the actual value, as shown in Fig. 7.5 and Table 7.2.

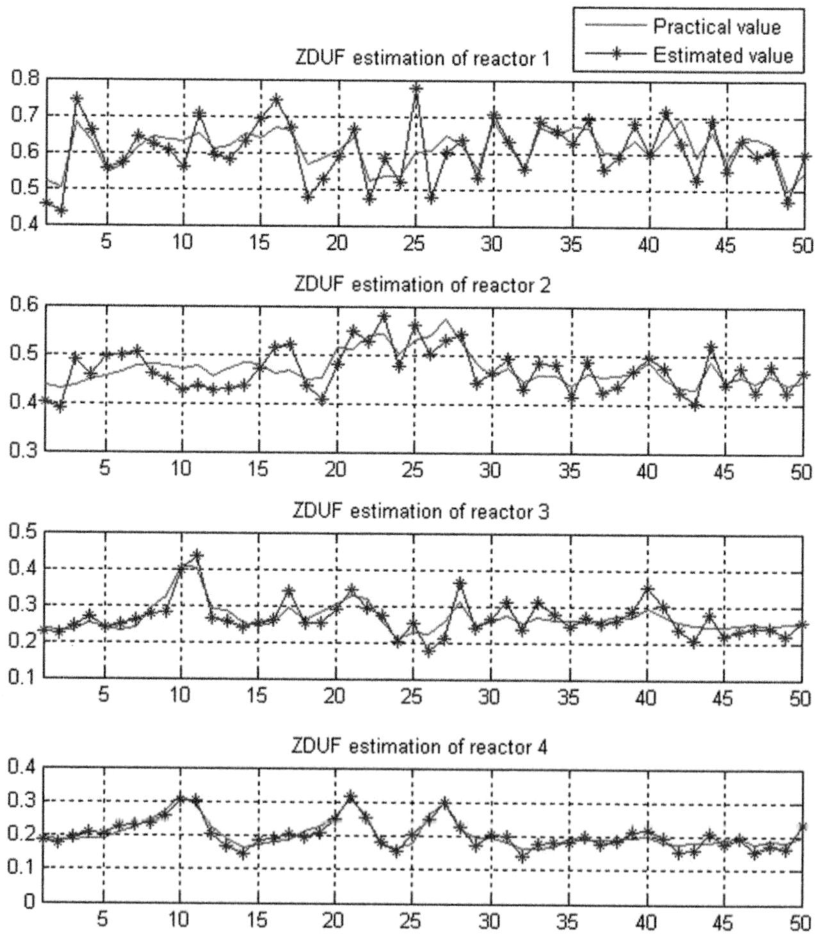

FIGURE 7.5 ZPUF estimation results of each reactor.

TABLE 7.2 Performance of the ZPUF estimation model.

Performance measure	Reactor 1	Reactor 2	Reactor 3	Reactor 4
Maximum relative error	29.01%	12.49%	20.37%	18.12%
Average relative error	6.56%	5.77%	7.99%	7.55%

7.2.2.2 *Rolling gradient optimization of ACP*

In every optimization step, the gradient optimization problem of ACP is solved to obtain the optimal setting value of CRR which is expressed as a vector $[\lambda_{Co}(1), \lambda_{Co}(2), \lambda_{Co}(3), \cdots, \lambda_{Co}(N)]$. However, in the formulation of the optimization problem, we considered the retention interval of a block of solution with volume V from flowing into the first reactor to overflowing the last reactor, as shown in Fig. 7.6. In practice, the solution flows continuously through ACP. The ZPUF of each reactor may change during this interval. In addition, the outlet cobalt ion concentrations of each reactor may not stay exactly at their optimal setting values. So in the solution of the GOACP problem, only $\lambda_{Co}(1)$ is used at the present optimization step, and $[\lambda_{Co}(2), \lambda_{Co}(3), \cdots, \lambda_{Co}(N)]$ needs to be recalculated at subsequent optimization steps. As shown in Fig. 7.7, the blue solid curve is the initially optimized decline curve obtained, and the dash curve is the practical decline curve. At the next optimization step, the CRR of reactor #2 is reoptimized, and so forth for reactor 3 and reactor 4.

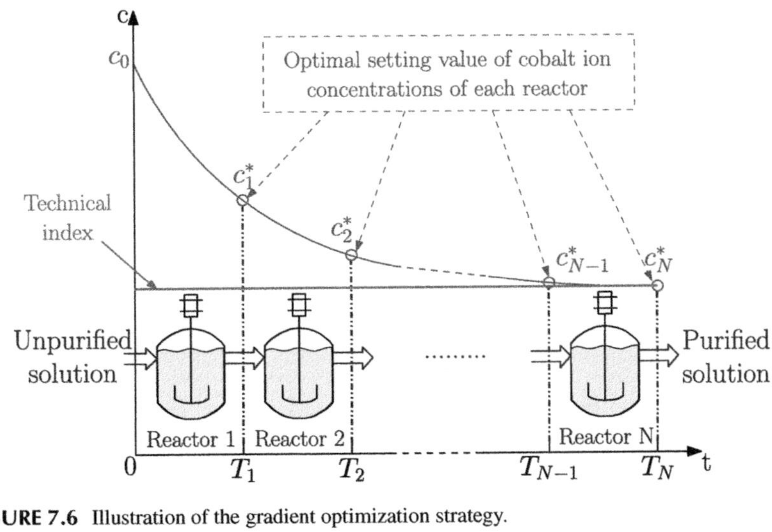

FIGURE 7.6 Illustration of the gradient optimization strategy.

For the above reasons, a rolling gradient optimization of ACP (RGOACP) is proposed. The steps of RGOACP include:

Step 1. Determine the optimization cycle.

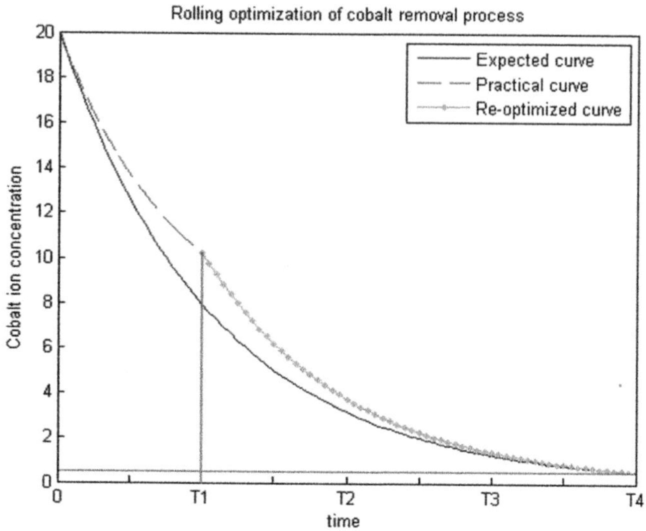

FIGURE 7.7 Rolling optimal setting of the CRR.

Step 2. At the beginning of the current optimization cycle, estimate the ZPUF of each reactor using the RBFNN. Assume the ZPUF of each reactor is constant during this optimization cycle.

Step 3. Optimize the CRRs of each reactor according to their ZPUFs, and record $\lambda_{Co}(1)$. Let $j = 2$.

Step 4. Then considering reactors #j to #N, calculate $[\lambda_{Co}(j), \lambda_{Co}(j+1), \cdots, \lambda_{Co}(N)]$, and record $\lambda_{Co}(j)$.

Step 5. Let $j = j + 1$. Repeat **Step 4** until $j = N$.

Step 6. If the current optimization cycle ends, start the next optimization cycle, and repeat **Step 2.**

7.3 Abnormal-state adjustment

The RGOACP framework provides an approach for the optimal control of ACP. However, this optimization strategy has its limitation. Fluctuations in the physicochemical properties of zinc concentrates, improper operation, and external disturbances will cause fluctuations in the operating state. Therefore, ACP cannot always stay in a normal state. When the process is in an abnormal state, it is more important to recover to a normal state through reasonable operations rather than to optimize zinc powder consumption. In addition, if the running state fluctuates severely, only adjusting the zinc powder dosage is insufficient to recover ACP to the normal state. Human operators gained rich experience during routine production. Their successful experience is sufficient to conduct reasonable operations under complex operating states. To detect the abnormal state, in this section, a data-driven online operating state monitoring method is first presented. Then, a CBR-based adjustment approach which utilizes human operators' successful experience for abnormal-state adjustment is introduced.

7.3.1 Data-driven online operating state monitoring

Data-driven online process monitoring is an approach to detect the type of current operating state. Multivariate statistical process monitoring (MSPM) is a widely applied data-driven online process monitoring method [11–14]. It recognizes the type of current operating state by observing the underlying mode of process variables [15].

In ACP, the process variables, including output states, inlet conditions, and reaction conditions, are highly correlated. Variation of the process variables is driven by some underlying events, e.g., change of raw materials and manipulated variables, maintenance, etc. [11]. These underlying events give rise to the change of working conditions, which can be reflected by the variation of process variables. The MSPM method detects the mode change of process variables by using multivariate statistical projection (MSP) methods [16]. These methods project the correlated raw process variables to a latent variable space. The latent variables are linearly uncorrelated, have lower dimension, and can be used to detect the operating state.

Principal component analysis (PCA) is a typical latent variable extraction method in process monitoring [17–19]. Denote $x \in R^m$ as the data sample including m process variables. If there exist N data samples representing the normal operating state, then a sample set containing the characteristic information of the normal operating state can be constructed as

$$X = [x_1\ x_2\ ...\ x_N]^T \in R^{N \times m},$$

where each row represents a data sample x_i^T.

The performance of PCA will be affected if the scales of the raw process variables are different. Therefore, the data sample set X is first scaled to zero mean and unit variance. The scaled data sample set is $(X - \mu)/D_\delta$. $\mu = [\mu_1\ \mu_2\ ...\ \mu_m]$ and $D_\delta = [\delta_1\ \delta_2\ ...\ \delta_m]$ are the mean value and standard deviation of the original data sample set, respectively. The covariance matrix of X is approximated by $\Sigma = X^T X/(N-1)$. The eigenvalues of the covariance matrix $\Lambda = diag[v_1\ v_2\ ...\ v_m]$, $v_{i-1} \geq v_i \geq 0\ (i = 2, 3, ..., m)$. Its orthogonal eigenvector is $p = [p_1\ p_2\ ...\ p_m]$, which satisfies $\Sigma = p\Lambda p^T$. If $\sum_{i=1}^{n_k} v_i / \sum_{i=1}^{m} v_i > \eta_v\ (1 \leq n_k \leq m)$, where η_v is a predefined contribution threshold of latent variables or principal components and n_k is the number of retained principal components, then X can be projected on the principal component subspace and the residual subspace as

$$X = pT + E, \tag{7.23}$$

where p is the principal loadings and $T = p^T X$ is the scores of n_k latent variables. The residual vector is therefore

$$E = X - Tp^T = (1 - pp^T)X. \tag{7.24}$$

After the construction of the PCA model, some statistical indices indicating the current operating state can be obtained, e.g., the Hotelling T^2 and squared prediction error (SPE) metrics [20]. Here, T^2 represents the deviation between the current operating state and the normal operating state and SPE is the residual or squared distance of the current observation from the model plane [11]:

$$T^2 = x_{new}^T p \Lambda^{-1} p^T x_{new}, \tag{7.25}$$

$$SPE = \|E_{new}\|^2 = \|(1 - pp^T)x_{new}\|^2. \tag{7.26}$$

In the building of the PCA model, only the data samples under the normal operating state are utilized. If the T^2 and SPE values of the current operating state exceed the upper limits [18], then it is regarded as an abnormal operating state. The upper limits of T^2 and SPE are [21]

$$T^2 \le \tau^2, \ \tau^2 = \frac{n_k(N^2 - 1)}{N(N - n_k)} F_{n_k, N-1, \alpha}, \tag{7.27}$$

$$SPE \le \gamma^2, \ \gamma^2 = \theta_1 \left(\frac{c_\alpha \sqrt{2\theta_2 h_0^2}}{\theta_1} + 1 + \frac{\theta_2 h_0 (h_0 - 1)}{\theta_1^2} \right)^{1/n_k}, \tag{7.28}$$

where $F_{n_k, N-1, \alpha}$ is an F distribution with n_k and $N - n_k$ degrees of freedom for a given significance level α, $\theta_i = \sum_{j=k+1}^{m} \lambda_j^i$ $(i = 1, 2, 3)$, $h_0 = 1 - (2\theta_1\theta_3)/(3\theta_2^2)$, and c_α is the normal deviate corresponding to the upper $(1 - \alpha)$ percentile.

In order to evaluate the detection performance of T^2 and SPE, the PCA model and the upper limits of T^2 and SPE are obtained using 100 data samples under the normal operating state. A number of 50 other data samples were used for testing. The test results are shown in Figs. 7.8 and 7.9. Samples exceeding the upper limits of T^2 and SPE are considered as abnormal operating states. It can be observed from the test result that:

(i) Samples detected as abnormal operating states have larger modeling errors.
(ii) SPE mainly reflects the stochastic disturbance forced on the process; it has a more intense distribution and clear variation trajectory than T^2; and
(iii) T^2 mainly reflects the variation of the essential process dynamics.

7.3.2 CBR-based adjustment under abnormal-state conditions

If the operating state of ACP is detected as abnormal, a reasonable determination of manipulating variables is needed to guarantee the satisfaction of technical

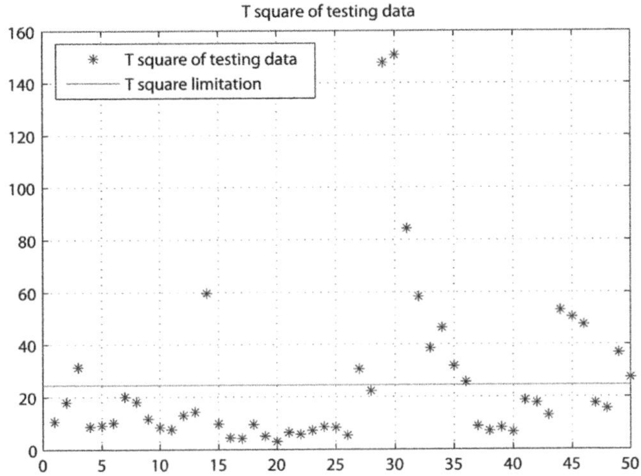

FIGURE 7.8 T^2 of test sample.

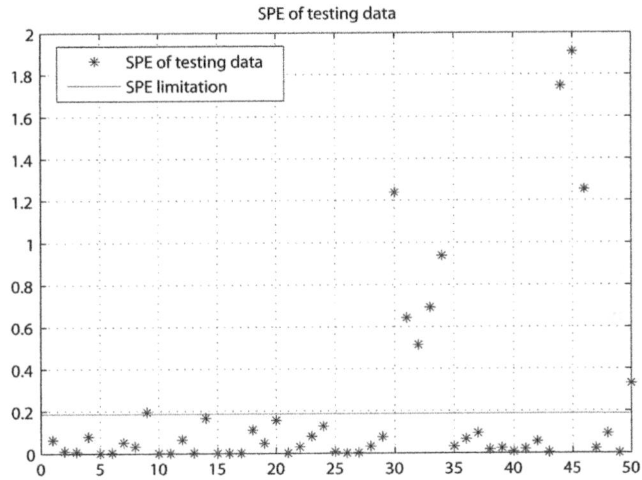

FIGURE 7.9 SPE of test sample.

requirements. In this section, a CBR-based controller for abnormal states is presented. CBR is an automated problem-solver which contains cases of past successful manipulations. Each case in the case base is composed of two parts [22]:

(i) Case feature: attributes describing the operating state,
(ii) Case solution: values of manipulated variables.

CBR retrieves most similar cases from a case base. The similarity is measured by comparing the output states, inlet conditions, and reaction conditions between the current process and cases in the base. The cases with the higher similarity are reused to produce the setting values of manipulated variables.

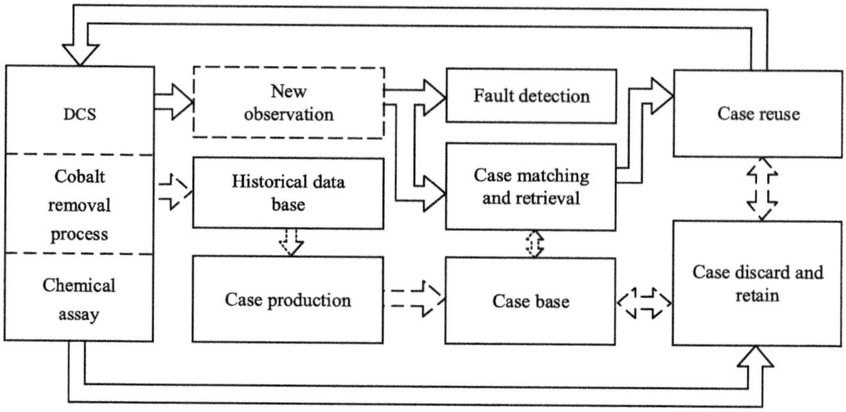

FIGURE 7.10 Structure of the CBR algorithm.

As shown in Fig. 7.10, the CBR-based controller is composed of five parts: (i) case production; (ii) case matching and retrieval; (iii) case reuse; (iv) case discard and retain; and (v) fault detection. To construct the case base, historical data of process variables and chemical assay results are used to form each case in the case production unit. These cases are stored in the case base. When new observations are obtained, it is first fed to the fault detection unit. If a fault is detected, then an alarm is sent to the operators. If the process is just under abnormal-state conditions, then the algorithm searches similar cases from the case base. These similar cases are then reused to provide reasonable adjustments. The detailed working procedure of these units is given below:

(i) Case production: The process variables and test values of outlet cobalt ion concentrations of each reactor are routinely collected by DCS and chemical assay. These records can be organized and stored in a historical database. However, the case base is different from the historical database. It collects successful operation cases under representative situations. Each case is composed of case feature and case solution. Case feature is a set of process variables reflecting the operating state. Case solution includes setting values of manipulating variables given which have been proved reasonable in previous operations. The feature and solution of the case are shown in Table 7.3.

(ii) Case matching and retrieval: Denote the feature of the current process as $V_c = [v_1, v_2, ..., v_{15}]$. Then the similarity between the current situation and case $C_k = c_{k,i}$ is

$$SIM(V_c, C_k) = \frac{\sum_{i=1}^{15} \omega_i sim(v_i, c_{k,i})}{\sum_{i=1}^{15} \omega_i}, \tag{7.29}$$

TABLE 7.3 Case presentation.

	Physical meaning	Symbol
Case feature	Flux of zinc sulfate solution	c_1
	Flux of underflow	c_2
	Dosage of arsenic trioxide	c_3
	Specific gravity of thickener underflow	c_4
	Flux of spent acid	c_5, c_6, c_7
	Zinc powder dosage of four reactors	c_8, c_9, c_{10}, c_{11}
	ORP of four reactors	$c_{12}, c_{13}, c_{14}, c_{15}$
Case solution	Dosage of arsenic trioxide	s_1
	Flux of spent acid	s_2
	Zinc powder dosage of four reactors	s_3, s_4, s_5, s_6
	ORP of four reactors	s_7, s_8, s_9, s_{10}

where $k = 1, 2, ..., K$, ω_i is the weight of feature i determined based on experience, and $sim(v_i, c_{k,i})$ is the similarity of the ith feature between V_c and C_k; $sim(v_i, c_{k,i})$ is defined as

$$sim(v_i, c_{k,i}) = 1 - \frac{|v_i - c_{k,i}|}{max(v_i, c_{k,i})}. \tag{7.30}$$

After the calculation of similarity, the cases that have a similarity larger than a threshold θ are retrieved to derive a reasonable solution for the current process.

(iii) Case reuse: If more than one case is retrieved, the cases with the minimum total zinc powder consumption are selected for obtaining the final solution. The two cases, denoted as C^1_{min} and C^2_{min}, which have the minimum zinc powder consumption or have the most positive ORP will be chosen to generate the setting value for the control variables:

$$C_s = \frac{SIM(V_c, C^1_{min}) \times C^1_{min} + SIM(V_c, C^2_{min}) \times C^2_{min}}{SIM(V_c, C^1_{min}) + SIM(V_c, C^2_{min})}, \tag{7.31}$$

where C_s is the output solution.

(iv) Case discard and retain: The performance of C_s is evaluated after it has been forced on the process. If the outlet cobalt concentrations of each reactor are within desired ranges and the fluctuations of the CRRs of each reactor are small, then C_s is considered to be a satisfying solution for the current situation and retained as a new case. Otherwise, it will be discarded.

(v) Fault detection: In most cases, abnormal operating states can be regulated to the normal state by adjusting manipulating variables. However, when there exists a fault in ACP, e.g., equipment failure, only adjusting manipulating variables may not help. Therefore, detection algorithms for typical faults of ACP are used to analyze the value and trend of related process variables. If a fault is detected, a fault alarm is triggered to remind the operators.

7.4 Results

To test the intelligent optimal setting control framework, an industry experiment was carried out. An experiment interval of 24 hours was selected to illustrate the performance of the framework, which was from 9:00 a.m., September 20, 2012, to 8:00 a.m., September 21, 2012.

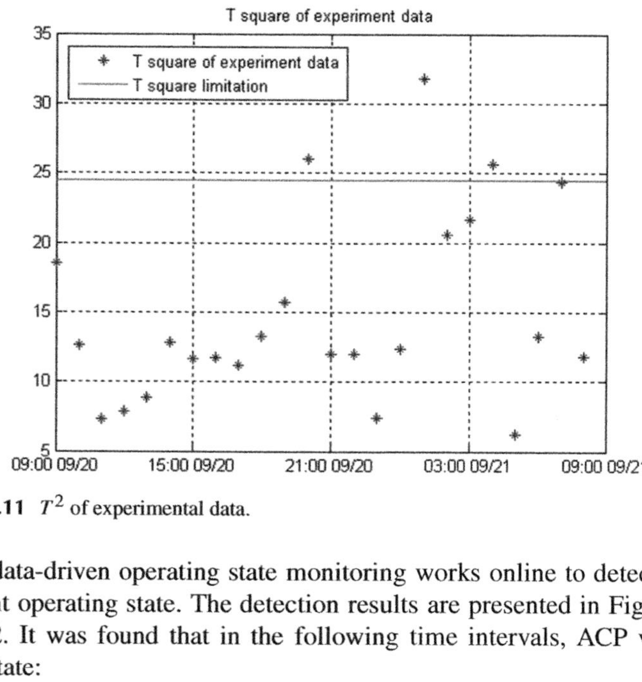

FIGURE 7.11 T^2 of experimental data.

The data-driven operating state monitoring works online to detect the type of current operating state. The detection results are presented in Fig. 7.11 and Fig. 7.12. It was found that in the following time intervals, ACP was in the normal state:

(i) [9:00 a.m., September 20, to 7:00 p.m., September 20],
(ii) [9:00 p.m., September 20, to 12:00 a.m., September 21],
(iii) [2:00 a.m., September 21, to 3:00 a.m., September 21],
(iv) [5:00 a.m., September 21, to 8:00 a.m., September 21].

In the above intervals, the RGOACP was used to optimize the total zinc powder consumption. The remaining intervals were in an abnormal state, and the CBR-based controller was used to obtain the operating variables:

(i) At 8:00 p.m., September 20, the flow rate of the inlet zinc sulfate solution was increased from 240 m^3/h to 300 m^3/h. As a result, the ORP of reactor #1 was increased, which indicates a decrease of reaction rate. This fluctuation was detected by T^2 and SPE simultaneously. As a response, the

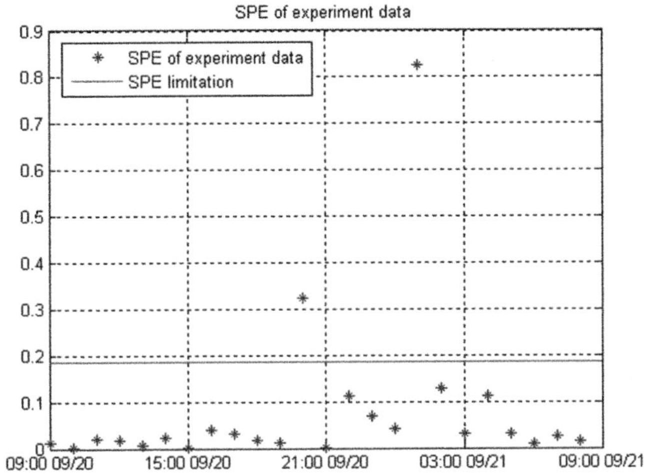

FIGURE 7.12 SPE of experimental data.

zinc dust dosage of reactor #1 was increased from 0.3 kg/m^3 to 0.4 kg/m^3. This adjustment is reasonable. The increase of flow rate will decrease the retention time of zinc sulfate solution in the reactor. Therefore, more zinc powder is added to increase the collision probability between cobalt ions and zinc powder particles.

(ii) At 1:00 a.m., September 21, an abnormal state was detected by both T^2 and SPE. The fault detection unit found that the zinc powder feeding device was blocked. In this abnormal state, increasing the zinc powder dosage is useless. Therefore, an alarm was sent to the operator for clearing the fault.

(iii) At 4:00 a.m., September 21, the inlet copper ion concentration increased from 120 mg/L to 200 mg/L. In this case, more zinc powder and arsenic trioxide were added to eliminate the effect of the increase of copper ion concentration.

The setting values of ORP and dosages of zinc powder and arsenic trioxide during the experiment are shown in Fig. 7.13, Fig. 7.14, and Fig. 7.15. The CRR and outlet cobalt ion concentrations of each reactor are shown in Fig. 7.16 and Fig. 7.17.

In order to evaluate the performance of the proposed control framework, the daily average effluent cobalt ion concentration (i.e., the outlet cobalt ion concentration of the last reactor) and total zinc powder consumption during the experiment were calculated and compared with those of the remaining days in September 2012, as shown in Table 7.4. By analyzing the experimental results, it can be observed that:

(i) The daily average total zinc powder consumption during the experiment was 0.6518 kg/m^3. This is less than the average zinc powder consumption on the remaining days of September 2012, which was 0.7652 kg/m^3.

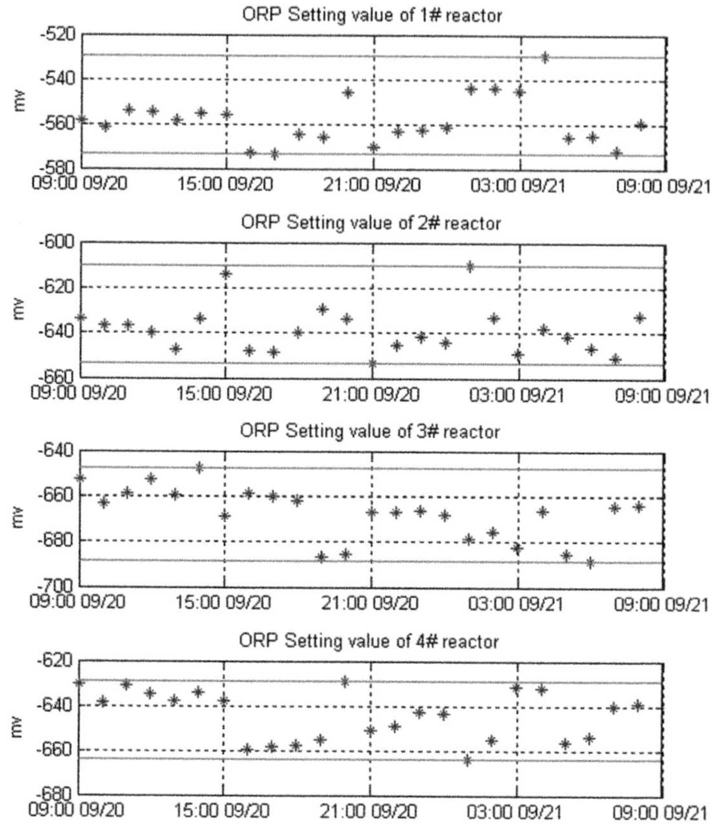

FIGURE 7.13 ORP setting value.

(ii) Compared with 86.96% of the remaining days in September 2012, the daily average zinc powder consumption during the experiment was lower.

(iii) The daily average effluent cobalt ion concentration during the experiment was 0.3792 mg/L. This is nearly equal to the average effluent cobalt ion concentration of the other days in September 2012, and satisfies the technical requirement, which is 0.5 mg/L.

These conclusions indicate that by using the proposed control strategy, the zinc powder consumption can be reduced while the effluent cobalt ion concentration can still meet the technical requirement.

On the other hand, the reduction in zinc powder consumption brought about by the proposed framework is less compared that brought about by process improvements. At the end of 2011, the zinc hydrometallurgy plant has conducted a technical reform. For comparison, statistical data of April 2011 and September 2012 were analyzed (Fig. 7.18). It can be observed that the total zinc powder consumption has decreased significantly. The technical root of the improvement

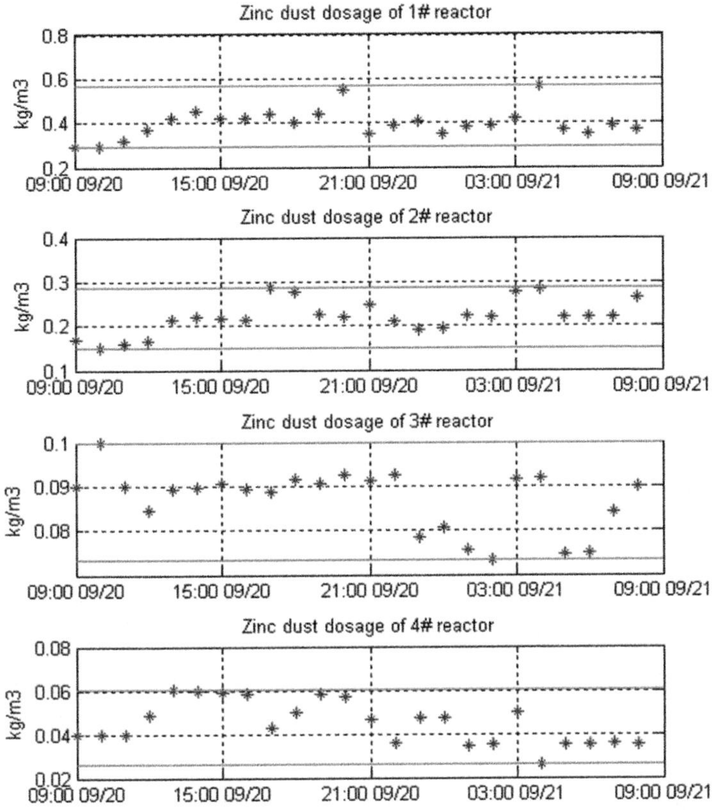

FIGURE 7.14 Zinc powder dosage setting value.

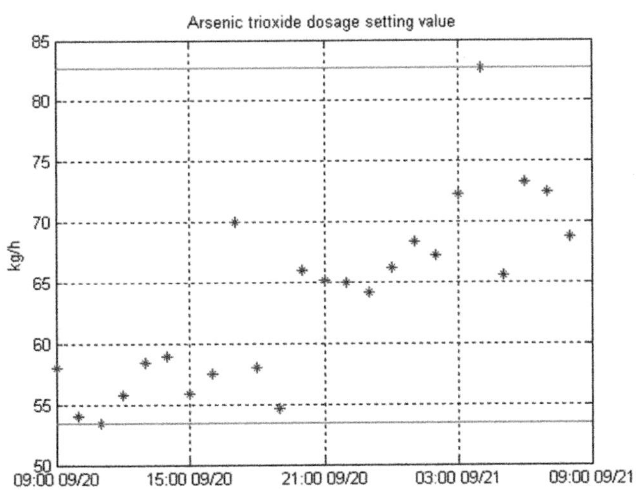

FIGURE 7.15 Arsenic trioxide dosage setting value.

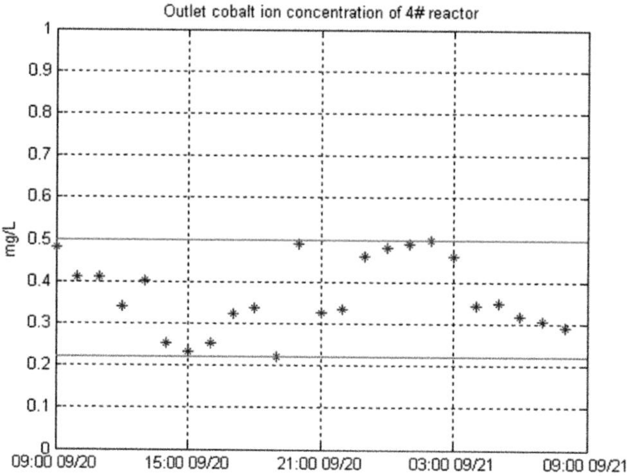

FIGURE 7.16 Effluent cobalt ion concentration.

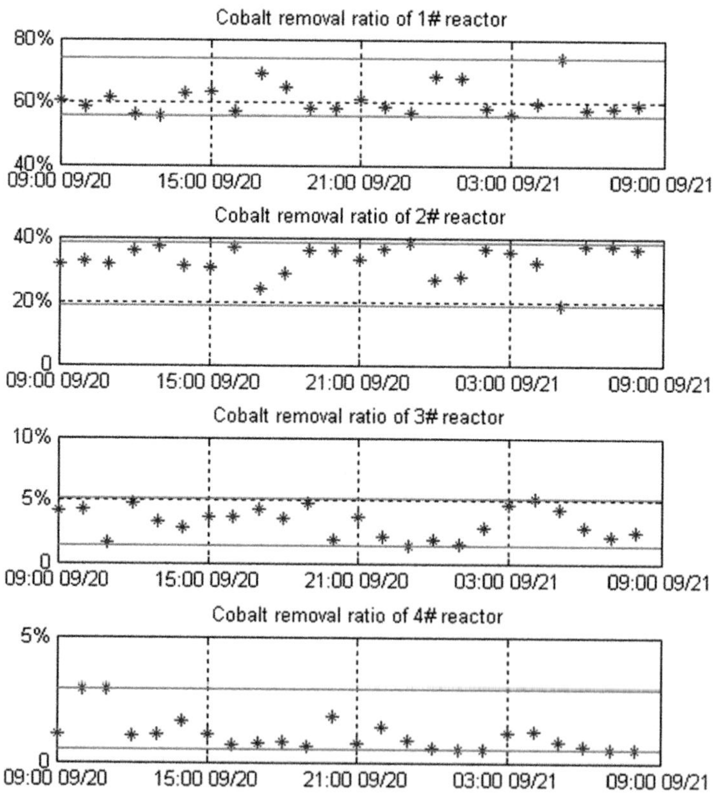

FIGURE 7.17 Cobalt removal ratio of each reactor.

TABLE 7.4 Comparison of average zinc powder consumption and effluent cobalt ion concentration.

Date	Zinc dust consumption (kg/m^3)	Effluent cobalt ion concentration (mg/L)	Date	Zinc dust consumption (kg/m^3)	Effluent cobalt ion concentration (mg/L)
Sep 19 to 25	0.6518	0.3792	Sep 12	0.7936	0.4986
Sep 1	0.6515	0.3328	Sep 13	0.8311	0.4857
Sep 12	0.6000	0.3700	Sep 14	0.9012	0.4367
Sep 13	0.6292	0.4586	Sep 15	0.7474	0.3329
Sep 14	0.7022	0.4143	Sep 16	0.9485	0.4029
Sep 15	0.7548	0.2657	Sep 17	0.8967	0.3543
Sep 16	0.7724	0.2971	Sep 18	0.7881	0.3314
Sep 17	0.7794	0.3543	Sep 26	1.0810	0.4515
Sep 18	0.7539	0.4014	Sep 27	0.7077	0.4871
Sep 19	0.7138	0.2471	Sep 28	0.6704	0.4571
Sep 10	0.6909	0.3229	Sep 29	0.7324	0.3771
Sep 11	0.7624	0.4171	Sep 30	0.6919	0.3114

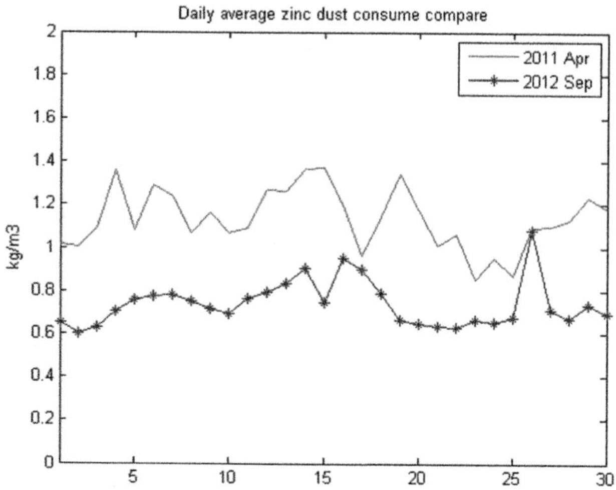

FIGURE 7.18 Comparison of daily average zinc powder consumption.

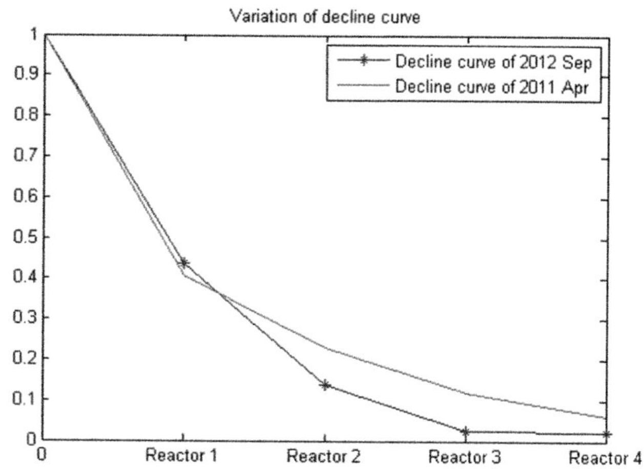

FIGURE 7.19 Variation of the average cobalt ion concentration decline curve.

is the reduction of inlet copper ion concentration and flow rate of spent acid. In addition, a change of cobalt ion concentration decline gradient is also observed, which also contributes to the reduction of zinc powder consumption (Fig. 7.19). This phenomenon proves the effectiveness of the idea of gradient optimization.

References

[1] A. Nelson, W. Wang, G.P. Demopoulos, G. Houlachi, The removal of cobalt from zinc electrolyte by cementation: a critical review, Mineral Processing and Extractive Metallurgy Review 20 (1) (2000) 325–356.

[2] S. Kim, K.E. Kim, E.K. Park, S.W. Song, S. Jung, Estimation methods for efficiency of additive in removing impurity in hydrometallurgical purification process, Hydrometallurgy 89 (2007) 242–252.

[3] B. Sun, W. Gui, Y. Wang, C. Yang, Intelligent optimal setting control of a cobalt removal process, Journal of Process Control 24 (5) (2014) 586–599.

[4] P. Zhou, T. Chai, H. Wang, Intelligent optimal-setting control for grinding circuits of mineral processing process, IEEE Transactions on Automation Science and Engineering 6 (4) (2009) 730–743.

[5] S. Dutta, B. Wierenga, A. Dalebout, Case-based reasoning systems: from automation to decision-aiding and stimulation, IEEE Transactions on Knowledge and Data Engineering 9 (6) (1997) 911–922.

[6] Ö. Yavuz, Y. Altunkaynak, F. Güzel, Removal of copper, nickel, cobalt and manganese from aqueous solution by kaolinite, Water Research 37 (4) (2003) 948–952.

[7] Q. Zha, Introduction to Electrode Process Kinetics, Science Press, Beijing, 2002.

[8] M. Wang, S.Y. Yang, S.J. Wu, F. Luo, A RBFNN approach for DoA estimation of ultra wideband antenna array, Neurocomputing 71 (4) (2008) 631–640.

[9] S. Chen, C.F.N. Colin, P.M. Grant, Orthogonal least squares learning algorithm for radial basis function networks, IEEE Transactions on Neural Networks 2 (2) (1991) 302–309.

[10] S.A. Iliyas, M. Elshafei, M.A. Habib, A.A. Adeniran, RBF neural network inferential sensor for process emission monitoring, Control Engineering Practice 21 (7) (2013) 962–970.

[11] J.F. MacGregor, A. Cinar, Monitoring, fault diagnosis, fault-tolerant control and optimization: data driven methods, Computers & Chemical Engineering 47 (2012) 111–120.

[12] R. Dunia, T.F. Edgar, T. Blevins, W. Wojsznis, Multistate analytics for continuous processes, Journal of Process Control 22 (2012) 1445–1456.

[13] C. Zhao, S. Mo, F. Gao, N. Lu, Y. Yao, Statistical analysis and online monitoring for handling multiphase batch processes with varying durations, Journal of Process Control 21 (2011) 817–829.

[14] C.F. Alcala, S.J. Qin, Reconstruction-based contribution for process monitoring, Automatica 45 (2009) 1593–1600.

[15] J.F. MacGregor, T. Kourti, Statistical process control of multivariate processes, Control Engineering Practice 3 (3) (1995) 403–414.

[16] T. Kourti, J.F. MacGregor, Process analysis, monitoring and diagnosis, using multivariate projection methods, Chemometrics and Intelligent Laboratory Systems 28 (1995) 3–21.

[17] S. Bouhouche, M. Yahi, J. Bast, Combined use of principal component analysis and self organisation map for condition monitoring in pickling process, Applied Soft Computing 11 (3) (2011) 3075–3082.

[18] N.A. Abd Majid, M.P. Taylor, J.J.J. Chen, B.R. Young, Multivariate statistical monitoring of the aluminium smelting process, Computers & Chemical Engineering 35 (2011) 2457–2468.

[19] M. He, C. Yang, W. Gui, Y. Ling, Performance recognition for sulphur flotation process based on froth texture unit distribution, Mathematical Problems in Engineering (2013) 530349.

[20] J.C. Gunther, J.S. Conner, D.E. Seborg, Process monitoring and quality variable prediction utilizing PLS in industrial fed-batch cell culture, Journal of Process Control 19 (2009) 914–921.

[21] S.J. Qin, Statistical process monitoring: basics and beyond, Journal of Chemometrics 17 (2003) 480–502.

[22] S. Petrovic, N. Mishra Nishikant, S. Sundar, A novel case based reasoning approach to radiotherapy planning, Expert Systems with Applications 38 (9) (2011) 10759–10769.

Chapter 8

Control of the cobalt removal process under multiple working conditions

Contents

8.1 Problem analysis

In Chapter 7, a "steady-state optimization, unsteady-state adjustment" optimal setting control strategy was established. An RGOACP economical optimization approach and a CBR-based adjustment approach were proposed. However, steady state and unsteady state are a rough classification of the operating states. Under steady-state conditions, there exist multiple working conditions, as discussed in Chapter 2. The integrated model proposed in Chapter 6 can provide satisfying modeling accuracy in most cases. However, model–plant mismatch inevitably happens due to reasons like equipment failure and external disturbances. In this situation, for a single reactor in ACP, the practical dynamics can be expressed as

$$\frac{dc}{dt} = \frac{f_{\text{in}}}{V}c_{\text{in}} - \frac{f_{\text{out}}}{V}c - A_{\beta}\exp(-\frac{E_{\text{e}} + 2\alpha\text{F}(e_{\text{orp}} - e_{\text{eq}})}{\text{R}T_c})c + \varepsilon_{\text{Mismatch}}, \quad (8.1)$$

where $\varepsilon_{\text{Mismatch}}$ denotes the mismatch, which is unknown [1][2][3][4][5][6][7]. Moreover, the reconfiguration of process flow and new types of raw materials may generate new working conditions. The insufficient number of data samples of new working conditions may lead to failure in accurately identifying model

Modeling, Optimization, and Control of Zinc Hydrometallurgical Purification Process
https://doi.org/10.1016/B978-0-12-819592-5.00019-3

parameters. As a result, the exact values of model parameters for the new working conditions are unknown, i.e., A_β, E_e, α, and e_{eq} are unknown.

For the "model–plant mismatch" situation, the controller should be able to account for the mismatch and guarantee the stability of the process. For the "unknown model parameters" situation, the controller should be able to provide control actions without using the process model. Therefore, to provide appropriate control decisions under the above situations, i.e., model–plant mismatch and unknown model parameters, two controller design approaches are presented in this chapter. A robust adaptive control [8] and an adaptive dynamic programming-based control [9] are designed for these two situations, respectively.

8.2 Robust adaptive control under model–plant mismatch

8.2.1 Nominal process model

Consider an ACP including N reactors. Then according to Eq. (8.1), the nominal dynamics of ACP can be described by

$$
\begin{aligned}
\frac{dc_1}{dt} &= \frac{f_{\text{in}}}{V}c_0 - \frac{f}{V}c_1 - A_{\beta 1}\exp(-\frac{E_{e1} + 2\alpha_1 \mathrm{F}(e_{\text{orp}1} - e_{\text{eq}1})}{\mathrm{R}T_c})c_1, \\
\frac{dc_2}{dt} &= \frac{f}{V}c_1 - \frac{f}{V}c_2 - A_{\beta 2}\exp(-\frac{E_{e2} + 2\alpha_2 \mathrm{F}(e_{\text{orp}2} - e_{\text{eq}2})}{\mathrm{R}T_c})c_2, \\
&\vdots \\
\frac{dc_N}{dt} &= \frac{f}{V}c_{N-1} - \frac{f}{V}c_N - A_{\beta N}\exp(-\frac{E_{eN} + 2\alpha_N \mathrm{F}(e_{\text{orp}N} - e_{\text{eq}N})}{\mathrm{R}T_c})c_N,
\end{aligned}
\tag{8.2}
$$

where $f = f_{\text{out}} = f_{\text{in}} + f_{\text{underflow}}$, where $f_{\text{underflow}}$ is the flow rate of underflow. According to Eq. (8.2), the nominal state space model of the process can be reformulated as

$$
\dot{\mathbf{x}} = \mathbf{s}_0 + \mathbf{A}\mathbf{x} - \mathbf{F}(\mathbf{x})\mathbf{g}(\mathbf{v}),
\tag{8.3}
$$

where $\mathbf{x} = [x_1, x_2, \cdots, x_N]^T = [c_1, c_2, \cdots, c_N]^T$ is the system state; the physical meaning of $\mathbf{x}$ is the outlet cobalt ion concentration of each reactor; $\mathbf{v} = [v_1, v_2, \cdots, v_N]^T = [e_{\text{orp}_1}, e_{\text{orp}_2}, \cdots, e_{\text{orp}_N}]^T$ is the ORP of each reactor; and $\mathbf{s}$, $\mathbf{A}$, $\mathbf{F}(\mathbf{x})$, and $\mathbf{g}(\mathbf{v})$ are:

$$
\mathbf{s}_0 = [\frac{f_{\text{in}}}{V}c_0, 0, \cdots, 0]^T,
\tag{8.4}
$$

$$\mathbf{A}=\begin{bmatrix} -a & 0 & 0 & 0 & \cdots & 0 \\ a & -a & 0 & 0 & \cdots & 0 \\ 0 & a & -a & 0 & \cdots & 0 \\ \vdots & \ddots & \ddots & \ddots & \ddots & \vdots \\ 0 & \cdots & 0 & a & -a & 0 \\ 0 & \cdots & 0 & 0 & a & -a \end{bmatrix}, \tag{8.5}$$

$$\mathbf{F}(\mathbf{x})=\begin{bmatrix} x_1 & 0 & \cdots & 0 \\ 0 & x_2 & \cdots & 0 \\ \vdots & \vdots & \ddots & \vdots \\ 0 & 0 & \cdots & x_N \end{bmatrix}, \tag{8.6}$$

$$\mathbf{g}(\mathbf{v})=\begin{bmatrix} g_1(v_1) \\ g_2(v_2) \\ \vdots \\ g_N(v_N) \end{bmatrix}, \tag{8.7}$$

where $a=\frac{f}{V}$, $g_i(v_i)=A_{\beta i}\exp(-\frac{E_{ei}+2\alpha_i \mathrm{F}(e_{\mathrm{orp}i}-e_{\mathrm{eq}i})}{\mathrm{R}T_c})$, $i=1,2,\cdots,N$.

8.2.2 Model–plant mismatch analysis

In practice, the dynamics of ACP are not always consistent. As discussed in Chapter 2, the dynamics of ACP are influenced by inlet and reaction conditions from the following aspects (Fig. 8.1):

(i) Model uncertainties: The kinetic model is built based on some assumptions which may not hold under certain working conditions. Therefore, there exist uncertainties in the kinetic model parameters which may not be compensated by the data-driven compensation model.

(ii) External disturbances: Measurement error and equipment failure could bring disturbances to ACP.

(iii) Reaction rate saturation: It can be observed from Eq. (8.1) that the ORP directly affects the reaction rate [6]. We have

$$k=A_\beta\exp(-\frac{E_\mathrm{e}+2\alpha\mathrm{F}(e_\mathrm{orp}-e_\mathrm{eq})}{\mathrm{R}T_c}).$$

However, the reaction rate cannot increase unlimitedly with the decrease of ORP. Saturation can occur between ORP and reaction rate. So in practice, ORPs of each reactor have their own adjusting ranges.

Therefore, for each reactor, the model parameters $\phi=[A_\beta, E_e, \alpha, e_\mathrm{eq}]$ exhibit uncertainties $\Delta\phi$; $\Delta\phi$ is unknown and depends on the inlet and reaction

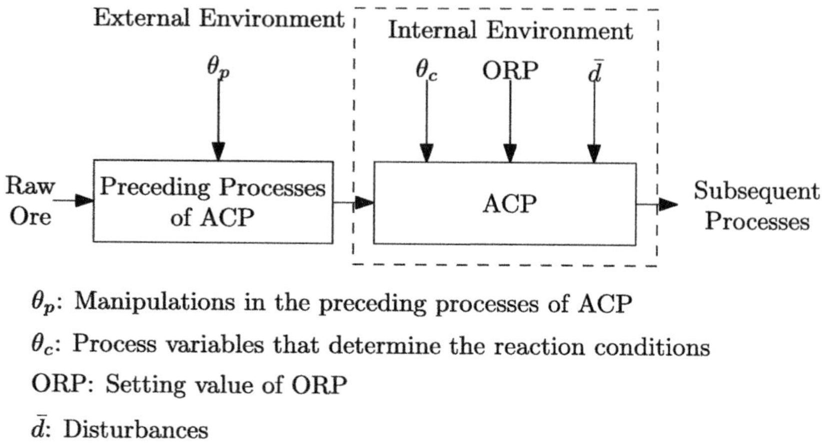

FIGURE 8.1 Influencing factors on the dynamics of ACP.

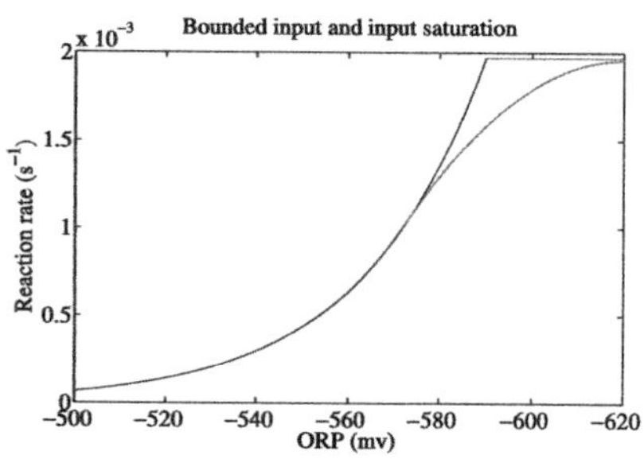

FIGURE 8.2 Reaction rate saturation phenomenon.

conditions. So $g(v)$ can be rewritten as

$$g(v) = \bar{g}(v) + d_g(\theta, t), \tag{8.8}$$

where $\theta = [\theta_p, \theta_c]^T$, $\bar{g}(v)$ is the nominal part of $g(v)$ by substituting the identified model parameters in $g(v)$, and $d_g(\theta, t)$ is the uncertain bias between $\bar{g}(v)$ and $g(v)$.

Due to the reaction rate saturation phenomenon, $\bar{g}(v)$ is saturated when v exceeds a certain value. As shown in Fig. 8.2, the saturation brings discontinuities, which hinders the application of controller design approaches. Considering the adjusting range of ORPs of each reactor and eliminating the discontinuities, a smooth function is used to approximate the saturated relation between ORP and reaction rate [10]:

$$g_a(v) = \begin{cases} \bar{a}v + bv^2 + c, & \text{if } v \geqslant v_M, \\ \bar{g}(v), & \text{if } v < v_M, \end{cases}$$

in which the parameters $\bar{a}$, b, c, and v_M are chosen according to $\bar{g}(v)$ (Fig. 8.2).

Then $\bar{g}(v)$ can be expressed as

$$\bar{g}(v) = g_a(v) + d_a(v), \tag{8.9}$$

where $d_a(v)$ is the bounded approximation error:

$$|d_a(v)| = |\bar{g}(v) - g_a(v)| \leqslant D_a. \tag{8.10}$$

Besides model uncertainties $d_g(\theta, t)$ and approximation error $d_a(v)$, the process dynamics can be influenced by external disturbances. Therefore, the process dynamics can be reformulated as

$$\begin{aligned}
\dot{x_0} &= h_0(\varphi, \theta_p, t), \\
\dot{x_1} &= a_0 x_0 - a x_1 - x_1 g_{a1}(v_1) + d_1(\theta, v_1, t), \\
\dot{x_i} &= a x_{i-1} - a x_i - x_i g_{ai}(v_i) + d_i(\theta, v_i, t),
\end{aligned} \tag{8.11}$$

in which $i = 2, 3, \cdots, N$, x_0 is the inlet cobalt ion concentration of the first reactor, φ denotes the physicochemical properties of zinc concentrates, $h_0(\varphi, \theta_p, t)$ denotes the dynamics of x_0 which are unknown, $a_0 = \frac{f_{\text{in}}}{V}$, and $d_i(\theta, v_i, t) = x_i(d_{g_i}(\theta, t) + d_{ai}(v_i)) + \bar{d}_i(t)$ $(i = 1, 2, 3, \cdots, N)$ accounts for the overall model–plant mismatch caused by model uncertainties, reaction rate saturation, and external disturbances.

8.2.3 Design of a robust adaptive tracking controller

To demonstrate the design process, we select reactor #1 as an example. As in practice, ORP is an intermediate variable, which can be controlled by adjusting the physical input or the zinc powder dosage (Fig. 8.3). It is considered as input when considering the cobalt ion concentration as the state. It is considered as state when considering zinc powder dosage as the input. To achieve closed-loop control of the cobalt ion concentration, zinc powder dosage is included in the controller design as control input.

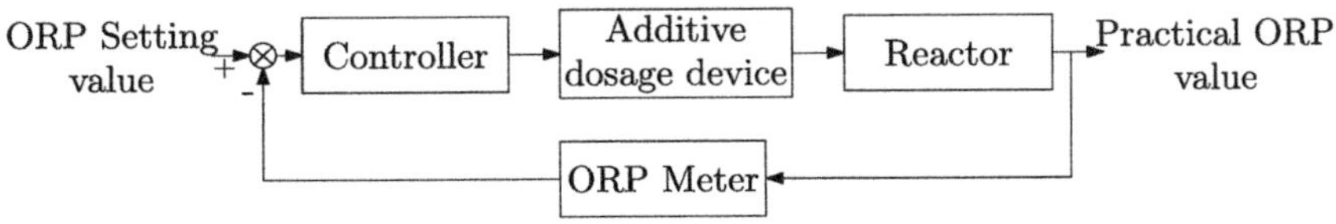

FIGURE 8.3 ORP-zinc powder dosage controller.

Before the development of the robust adaptive tracking controller, the following assumptions are made:

Assumption 1: ACP is input-to-state stable (ISS).
Assumption 2: Process variables of ACP are bounded.

To start with, the following system is formulated for the controller design of reactor #1:

$$\dot{x_0} = h_0(\varphi, \theta_p, t),$$
$$\dot{x_1} = a_0 x_0 - a x_1 - h_1 + d_1(t), \tag{8.12}$$
$$\dot{h}_1 = \omega,$$

where $h_1 = x_1 g_{a1}(v_1)$, ω is to be designed, and $d_1(t)$ is bounded by D_1 which is unknown. The control input v_1 is embedded in h_1; h_1 is a function of x_1 and v_1. For given h_1 and x_1, the value of x_1 is determined. In the following, we derive the control law by designing h_1.

Denote the following variables:

$$z_1 = x_1 - x_{r1},$$
$$z_2 = h_1 - \alpha_1, \tag{8.13}$$

where x_{r1} is the optimal setting value of x_1, z_1 is the error between x_1 and its setting value, and z_2 is introduced for h_1. Similarly to ω, α_1 is related with h_1 and needs to be designed.

As there exist unknown and time-varying variables in the process model and the control object is to obtain a stabilizing control law, a robust adaptive controller design approach is applied. The controller design is based on robust adaptive control and backstepping, which includes the following steps.

Step 1: Design a Lyapunov function,

$$V_{z1} = \frac{1}{2} z_1^2, \tag{8.14}$$

with its derivative

$$\begin{aligned} \dot{V}_{z1} &= z_1 \dot{z}_1 \\ &= z_1[a_0 x_0 - a x_1 - z_2 - \alpha_1 + d_1(t) - \dot{x}_{r1}]. \end{aligned} \tag{8.15}$$

If we denote the bounds of $d_1(t)$ and x_0 as D_1 and D_0, then $z_1 d_1(t) \leqslant |z_1| D_1$, $z_1 x_0 \leqslant |z_1| D_0$, and

$$\dot{V}_{z1} \leqslant z_1[-a x_1 - z_2 - \alpha_1 - \dot{x}_{r1}] + a_0 |z_1| D_0 + |z_1| D_1, \tag{8.16}$$

$$\alpha_1 = c_1 z_1 - a x_1 - \dot{x}_{r1} + a_0 sgn(z_1) \hat{x}_0 + sgn(z_1) \hat{D}_1, \tag{8.17}$$

with the unknown variable $\hat{x}_0$ adaptive with z_1 as follows:

$$\dot{\hat{x}}_0 = \eta_0 a_0 |z_1|, \tag{8.18}$$

where $\hat{x}_0$ is the estimation of D_0. Design another Lyapunov function,

$$V_1 = V_{z1} + \frac{1}{2\eta_0} \tilde{x}_0^2, \tag{8.19}$$

where $\tilde{x}_0 = D_0 - \hat{x}_0$. Then combine Eq. (8.16), Eq. (8.17), and Eq. (8.19) to obtain

$$\begin{aligned}\dot{V}_1 &= z_1\dot{z}_1 + \frac{1}{\eta_0}\tilde{x}_0\dot{\tilde{x}}_0 \\ &\leqslant -c_1 z_1^2 - z_1 z_2 + |z_1|\tilde{D}_1. \end{aligned} \tag{8.20}$$

Step 2: Design the following Lyapunov function:

$$V_{z2} = \frac{1}{2}z_2^2. \tag{8.21}$$

The derivative of V_{z2} is

$$\begin{aligned}\dot{V}_{z2} &= z_2\dot{z}_2 = z_2(\omega - \dot{\alpha}_1) \\ &= z_2\{\omega - [(c_1 - a)\dot{x}_1 + a_0 sgn(z_1)\dot{\hat{x}}_0 - c_1\dot{x}_{r1} - x_{r1}^{(2)} + sgn(z_1)\dot{\hat{D}}_1]\} \\ &= z_2\{\omega - [(c_1 - a)(a_0 x_0 - a x_1 - h_1 + d(t)) + a_0 sgn(z_1)\dot{\hat{x}}_0 \\ &\quad - c_1\dot{x}_{r1} - x_{r1}^{(2)} + sgn(z_1)\dot{\hat{D}}_1]\} \\ &\leqslant z_2\{\omega - [(c_1 - a)(a_0 x_0 - a x_1 - h_1 - sgn(z_2)sgn(c_1 - a)D_1) \\ &\quad + a_0 sgn(z_1)\dot{\hat{x}}_0 - c_1\dot{x}_{r1} - x_{r1}^{(2)} + sgn(z_1)\dot{\hat{D}}_1]\}. \end{aligned} \tag{8.22}$$

If

$$\begin{aligned}\omega &= -c_2 z_2 + z_1 + (c_1 - a)(a_0 x_0 - a x_1 - h_1) - sgn(z_2)|c_1 - a|\hat{D}_1 \\ &\quad + a_0 sgn(z_1)\dot{\hat{x}}_0 - c_1\dot{x}_{r1} - x_{r1}^{(2)} + sgn(z_1)\dot{\hat{D}}_1, \end{aligned} \tag{8.23}$$

with the estimation $\hat{D}_1$ adaptive with z_1 and z_2 as

$$\dot{\hat{D}}_1 = \eta_1(|z_1| + |z_2(c_1 - a)|), \tag{8.24}$$

then, for the Lyapunov function

$$V_2 = V_1 + V_{z2} + \frac{1}{2\eta_1}\tilde{D}_1^2, \tag{8.25}$$

where $\tilde{D}_1 = D_1 - \hat{D}_1$, the following is satisfied:

$$\dot{V}_2 = z_1\dot{z}_1 + z_2\dot{z}_2 + \frac{1}{\eta_0}\tilde{x}_0\dot{\tilde{x}}_0 + \frac{1}{\eta_1}\tilde{D}_1\dot{\tilde{D}}_1 \leqslant -c_1 z_1^2 - c_2 z_2^2. \tag{8.26}$$

This indicates that if using the above design, the outlet cobalt ion concentration can converge to its optimal setting value.

Theorem 1. *For reactor #1 of ACP, its outlet cobalt ion concentration is global asymptotically stable if*

$$v_1 = g_{a_1}^{-1}(h_1/x_1), \tag{8.27}$$

with

$$\begin{aligned}
\dot{h}_1 &= \omega, \\
\omega &= -c_2z_2 + z_1 + (c_1 - a)(a_0x_0 - ax_1 - h_1) - sgn(z_2)|c_1 - a|\hat{D}_1 \\
&\quad + a_0sgn(z_1)\dot{\hat{x}}_0 - c_1\dot{x}_{r1} - x_{r1}^{(2)} + sgn(z_1)\dot{\hat{D}}_1, \\
\dot{\hat{x}}_0 &= \eta_0a_0|z_1|, \\
\dot{\hat{D}}_1 &= \eta_1(|z_1| + |z_2(c_1 - a)|).
\end{aligned}$$

Proof. **Theorem 1** can be established from the analysis in ***Step 1*** and ***Step 2***, and is omitted here for brevity. □

Remark 1: After ***Step 1*** and ***Step 2***, the reference trajectory of ORP is obtained. The ORP reference trajectories of other reactors can be obtained following the same lines of reasoning. However, to achieve closed-loop control of the outlet cobalt ion concentration, an "ORP-zinc powder dosage" controller is still required.

Remark 2: The relationship between ORP and zinc powder is very complex. There is no existing model describing how the zinc powder dosage changes the value of ORP. In the following steps, an approximating system is first constructed. Then, a zinc powder dosage controller is designed.

Step 3: Formulate the "ORP-zinc powder dosage" system as

$$\dot{v}_1 = \zeta_1(x_0 - \mu_1u_1) + \sigma_1(x_1, \theta, v_1, t), \tag{8.28}$$

in which $\zeta_1(x_0 - \mu_1u_1)$ is the empirical part and $\sigma_1(x_1, \theta, v_1, t)$ is the bias between the empirical part and the real dynamics of v_1.

As the dynamics of v_1 is only partially known, we design the following approximation system [11]:

$$\begin{aligned}
\dot{\hat{v}}_1 &= f_v(u_1) + c_3sgv(z_3) + \hat{D}_3sat(z_3), \\
\dot{\hat{D}}_3 &= \eta_3|sgv(z_3)|,
\end{aligned} \tag{8.29}$$

where $f_v(u_1) = \zeta_1(x_0 - \mu_1u_1)$, c_3 is a positive design parameter, $z_3 = v_1 - \hat{v}_1$ denotes the approximation error, and $\hat{D}_3$ is the estimation of D_3, which is the bound of $\sigma_1(x_1, \theta, v_1, t)$.

Functions $sgv(z_3)$ and $sat(z_3)$ are defined as

$$sgv(z_3) = \begin{cases} z_3 - c_v, & \text{if } z_3 \geqslant c_v, \\ 0, & \text{if } |z_3| < c_v, \\ z_3 + c_v, & \text{if } z_3 \leqslant -c_v, \end{cases}$$

$$sat(z_3) = \begin{cases} sgn(z_3), & \text{if } |z_3| \geqslant c_v, \\ z_3/c_v, & \text{if } |z_3| < c_v, \end{cases}$$

where $c_v > 0$ is a parameter to be designed by the user.

Consider the following Lyapunov function [12]:

$$V_3 = \begin{cases} \frac{1}{2}(z_3 - c_v)^2 + \tilde{D}_3^2/(2\eta_3), & \text{if } z_3 \geqslant c_v, \\ \tilde{D}_3^2/(2\eta_3), & \text{if } |z_3| < c_v, \\ \frac{1}{2}(z_3 + c_v)^2 + \tilde{D}_3^2/(2\eta_3), & \text{if } z_3 \leqslant -c_v, \end{cases}$$

where $\tilde{D}_3 = D_3 - \hat{D}_3$ is the estimation error of D_3. Then the following theorem can be obtained.

Theorem 2. *For the "ORP-zinc powder dosage" system, by using the approximation system (8.29), the approximation error of* v_1 *can converge to* $[-c_v, c_v]$*, which can be chosen arbitrarily small by the user.*

Proof. The derivative of V_3 is

$$\dot{V}_3 = sgv(z_3)\dot{z}_3 + \frac{1}{\eta_3}\tilde{D}_3\dot{\tilde{D}}_3. \tag{8.30}$$

As $\dot{z}_3 = \dot{v}_1 - \dot{\hat{v}}_1 = f_v(u_1) + \sigma_1(x_1, \theta, v_1, t) - \dot{\hat{v}}_1$,

$$\begin{aligned} \dot{V}_3 &= sgv(z_3)(f_v(u_1) - \dot{\hat{v}}_1) + sgv(z_3)\sigma_1(x_1, \theta, v_1, t) + \frac{1}{\eta_3}\tilde{D}_3\dot{\tilde{D}}_3 \\ &\leqslant sgv(z_3)(f_v(u_1) - \dot{\hat{v}}_1) + |sgv(z_3)|D_3 + \frac{1}{\eta_3}\tilde{D}_3\dot{\tilde{D}}_3, \end{aligned} \tag{8.31}$$

we have

$$\begin{aligned} \dot{V}_3 &\leqslant -c_3 sgv(z_3)^2 - sgv(z_3)sgn(z_3)\hat{D}_3 + |sgv(z_3)|D_3 \\ &\quad - \frac{1}{\eta_3}\tilde{D}_3\dot{\hat{D}}_3 + sgv(z_3)[sgn(z_3)\hat{D}_3 - sat(z_3)\hat{D}_3] \\ &\leqslant -c_3 sgv(z_3)^2 + \frac{1}{\eta_3}\tilde{D}_3(\eta_3|sgv(z_3)| - \dot{\hat{D}}_3) \\ &\quad + sgv(z_3)[sgn(z_3)\hat{D}_3 - sat(z_3)\hat{D}_3]. \end{aligned} \tag{8.32}$$

From the definition, it can be observed that $sgv(z_3)[sgn(z_3)\hat{D}_3 - sat(z_3)\hat{D}_3] = 0$; therefore,

$$\dot{V}_3 \leqslant -c_3 sgv(z_3)^2. \tag{8.33}$$

This indicates that the approximation error z_3 asymptotically converges to ϵ, where $\epsilon = \{z_3 : |z_3| \leqslant c_v\}$ [12]. In addition, the estimation error of D_3 asymptotically converges to zero. □

Step 4: The tracking error of ORP is

$$\begin{aligned} z_5 &= v_{r1} - v_1 \\ &= v_{r1} - \hat{v}_1 + \hat{v}_1 - v_1 \\ &= z_4 - z_3, \end{aligned} \tag{8.34}$$

where $z_4 = v_{r1} - \hat{v}_1$ with the following derivative:

$$\begin{aligned} \dot{z}_4 &= \dot{v}_{r1} - \dot{\hat{v}}_1 \\ &= \dot{v}_{r1} - f_v(u_1) - c_3 sgv(z_3) - \hat{D}_3 sat(z_3) \\ &= \zeta_1 \mu_1 u_1 - \zeta_1 x_0 - c_3 sgv(z_3) - \hat{D}_3 sat(z_3) + \dot{v}_{r1}. \end{aligned} \tag{8.35}$$

If

$$u_1 = \frac{1}{\zeta_1 \mu_1}[\zeta_1 x_0 + c_3 sgv(z_3) + \hat{D}_3 sat(z_3) - \dot{v}_{r1} - c_4 z_4], \tag{8.36}$$

then

$$\dot{z}_4 = -c_4 z_4. \tag{8.37}$$

This indicates that z_4 converges asymptotically to zero.

According to ***Step 3***, z_3 asymptotically converges to $\epsilon = \{z_3 : |z_3| \leqslant c_v\}$. Therefore, z_5, which equals $(z_4 - z_3)$, also asymptotically converges to $\epsilon = \{z_5 : |z_5| \leqslant c_v\}$.

Remark 3: In ***Step 3*** and ***Step 4***, the zinc powder dosage control law is derived which enables the closed-loop control of ACP. Among the design parameters, increasing c_1, c_2, c_3, and c_4 or decreasing c_v can decrease the tracking error. However, on the other hand, increasing c_1, c_2, c_3, and c_4 will increase the control energy. Increasing η_1, η_2, η_3, and η_4 can reduce the influence of the initial estimation error.

8.2.4 Control performance analysis

In order to test the tracking performance of the robust adaptive controller, data samples collected from the site were utilized to conduct an experimental study. The interval of the experimental study is 24 hours. In the experiment, the ACP is simulated using a process model. To simulate the real situation, model parameter uncertainties, variation of inlet conditions, and external disturbances, which are placed on reactor #1 (Fig. 8.4), are included in the simulation. The nominal model used in controller design is different with the simulation model.

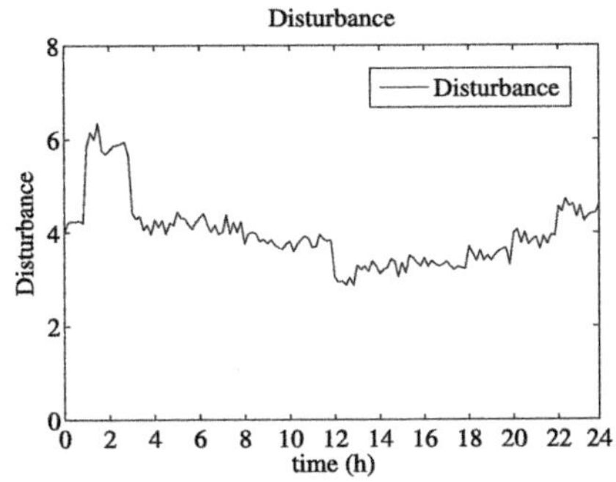

FIGURE 8.4 External disturbance acting on reactor #1.

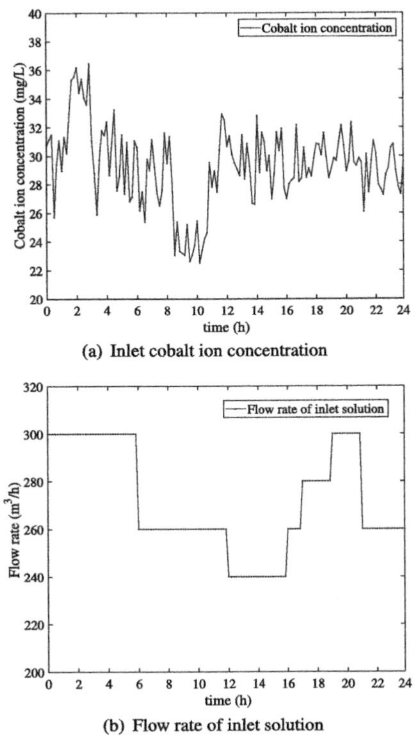

FIGURE 8.5 Variation of inlet conditions.

The variation of inlet conditions is shown in Fig. 8.5. The outlet cobalt ion concentrations of each reactor are precalculated using the RGOACP approach and shown in Fig. 8.6.

The tracking performance of the controller is shown in Fig. 8.7 and Fig. 8.8. It can be observed from the results that:

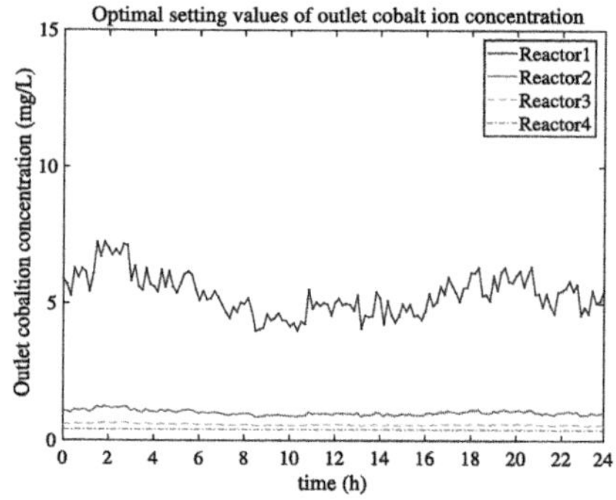

FIGURE 8.6 Optimal values of outlet cobalt ion concentrations of each reactor.

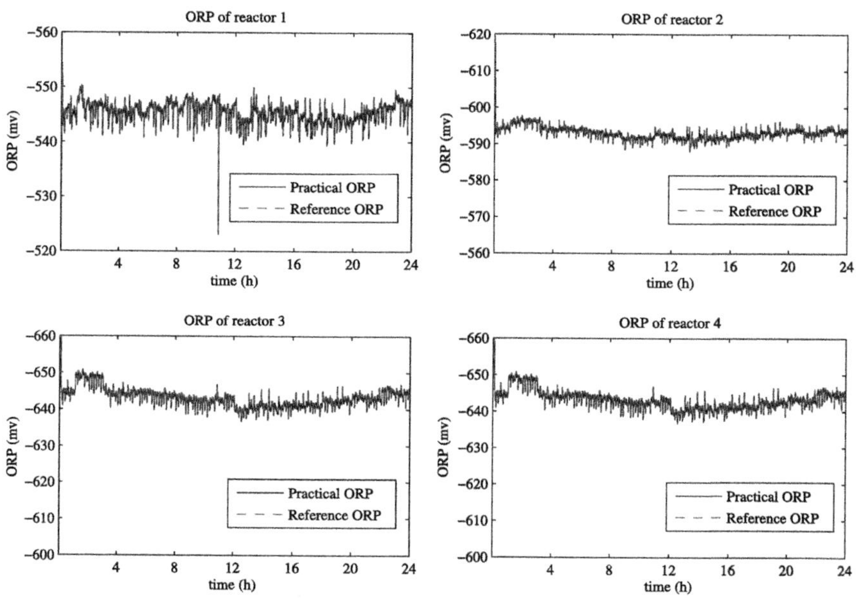

FIGURE 8.7 ORP tracking performance.

(i) The outlet cobalt ion concentration of the last reactor is always lower than the technical threshold.

(ii) For each reactor, by using the controller, the cobalt ion concentration and ORP can track their trajectories.

(iii) The fluctuation in the outlet cobalt ion concentration decreases along the reactors, which indicates that the robust adaptive controller attenuates the effects of model–plant mismatch.

Therefore, the robust adaptive control can be used as an alternative when there exists model–plant mismatch in ACP.

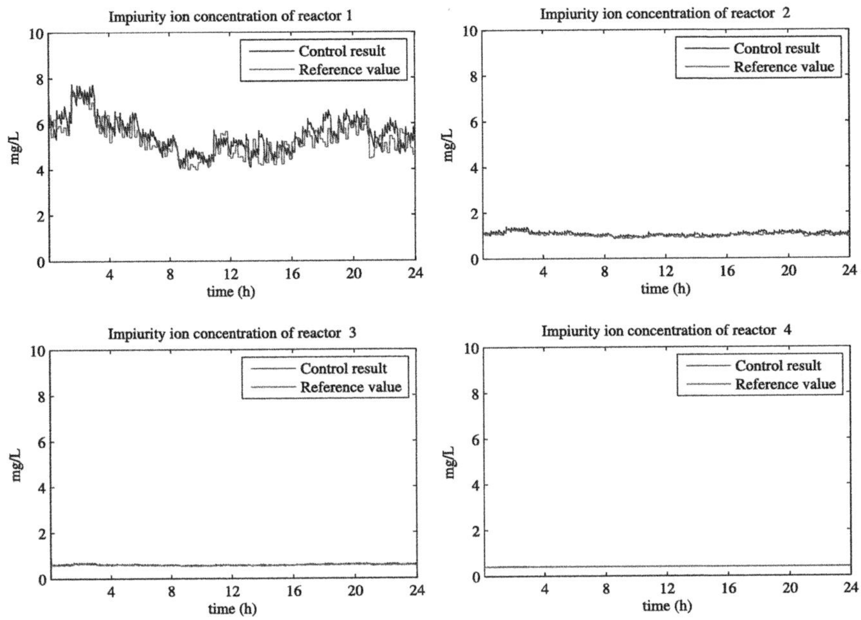

FIGURE 8.8 Outlet cobalt ion concentration tracking performance.

8.3 Adaptive dynamic programming for working conditions with unknown model parameters

In Section 8.2, a robust adaptive controller was designed to handle the model–plant mismatch situation [8]. However, in practice, new working conditions may emerge. The data samples associated with new working conditions are usually insufficient to identify the model parameters and train the data-driven compensation model. Therefore, under these working conditions, the model parameters are unknown. Model-based control approaches cannot be directly applied. A controller design approach which does not rely on the process model is required. In this section, we demonstrate the application of a model-free controller design approach for ACP.

8.3.1 Problem formulation

Consider the ith reactor ($i = 1, 2, \cdots, N$) with the following dynamics [6]:

$$
\begin{aligned}
\dot{c}_i &= \frac{f_{i-1}}{V} c_{i-1} - \frac{f_i}{V} c_i - A_\beta e^{-\frac{E_e + 2\alpha \mathrm{F}(v_i - v_{eq})}{RT}} c_i, \\
\dot{v_i} &= \zeta_i (c_{i-1} - \mu_i u_i) + \sigma_i (c_i, \theta, v_i, t).
\end{aligned}
$$

Denote c_i^* as the optimal setting value of c_i and

$$c_i' = c_i - c_i^*.$$

Then

$$\begin{aligned}\dot{c}_i' &= \frac{f_{i-1}}{V}(c_{i-1}' + c_{i-1}^*) - \frac{f_i}{V}(c_i' + c_i^*) - w(v_i', \theta_i')(c_i' + c_i^*),\\ \dot{v}_i &= \sigma_i(c_i', c_i^*, \theta, v_i, t) + \zeta_i(c_{i-1}' + c_{i-1}^*) - \zeta_i \mu_i u_i,\end{aligned}$$

where $i = 1, 2, \cdots, N$,

$$w(v, \theta') = A_\beta e^{-\frac{E_e + 2\alpha F(v - v_{eq})}{RT}},$$

and $\theta' = [A_s, E_e, \alpha, v_{eq}]$.

Denote

$$\mathbf{x} = [x_1 \quad x_2 \quad \cdots \quad x_n]^{\mathrm{T}} = [\mathbf{x}_{\mathrm{c}} \quad \mathbf{x}_{\mathrm{orp}}]^{\mathrm{T}} \quad \text{and} \quad \mathbf{u} = [u_1 \quad u_2 \quad \cdots \quad u_m]^{\mathrm{T}},$$

where $\mathbf{x}_{\mathrm{c}} = [c_1' \quad c_2' \quad \cdots \quad c_N']$, $\mathbf{x}_{\mathrm{orp}} = [v_1 \quad v_2 \quad \cdots \quad v_N]$, $n = 2N$, and $m = N$. Then

$$\dot{\mathbf{x}} = \mathbf{f}(\mathbf{x}) + \mathbf{g}(\mathbf{x})\mathbf{u}, \tag{8.38}$$

where $\mathbf{f}(\mathbf{x}) \in \mathbb{R}^{n\times 1}$, $\mathbf{g}(\mathbf{x}) \in \mathbb{R}^{n\times m}$, and

$$\mathbf{f}(\mathbf{x}) = \begin{bmatrix} \frac{f_0}{V}c_0 - \frac{f_1}{V}(x_1 + c_1^*) - \\ w(v_1, \theta_1')(x_1 + c_1^*) \\ \frac{f_1}{V}(x_1 + c_1^*) - \frac{f_2}{V}(x_2 + c_2^*) - \\ w(v_2, \theta_2')(x_2 + c_2^*) \\ \vdots \\ \frac{f_{N-1}}{V}(x_{N-1} + c_{N-1}^*) - \frac{f_N}{V}(x_N + c_N^*) \\ -w(v_N, \theta_N')(x_N + c_N^*) \\ \sigma_1 + \zeta_1 c_0 \\ \sigma_2 + \zeta_1(x_1 + c_1^*) \\ \vdots \\ \sigma_N + \zeta_N(x_N + c_N^*) \end{bmatrix},$$

$$\mathbf{g}(\mathbf{x}) = \begin{bmatrix} 0 & 0 & \dots & 0 \\ 0 & 0 & \dots & 0 \\ \vdots & \vdots & \ddots & \vdots \\ 0 & 0 & \dots & 0 \\ -\zeta_1\mu_1 & 0 & \dots & 0 \\ 0 & -\zeta_2\mu_2 & \dots & 0 \\ \vdots & \vdots & \ddots & \vdots \\ 0 & 0 & \dots & -\zeta_N\mu_N \end{bmatrix}.$$

Generally, for system (8.38), the control objective is to minimize

$$J = \int_t^\infty [Q(\mathbf{x}(\tau)) + W(\mathbf{u}(\tau))]d\tau. \tag{8.39}$$

The solution to the optimal control problem is $\mathbf{u}^*(\mathbf{x})$; $Q(\mathbf{x})$ and $W(\mathbf{u})$ are positive definite, which are usually chosen as $Q(\mathbf{x}) = \mathbf{x}^T\mathbf{Q}\mathbf{x}$ and $W(\mathbf{u}) = \mathbf{u}^T\mathbf{R}\mathbf{u}$; $\mathbf{Q} \in \mathbb{R}^{n\times n}$ and $\mathbf{R} \in \mathbb{R}^{m\times m}$ are diagonal matrices with their diagonal elements all positive.

If there exist physical limits on the input variable constraints, the performance index needs to be redesigned to include the input constraints. A general way to handle the input constraint is using the nonquadratic integrand in the performance index [13]:

$$W(\mathbf{u}) = 2\int_{\mathbf{0}}^{\mathbf{u}} (\mathbf{H}^{-1}(\mathbf{v}))^T\mathbf{R}d\mathbf{v}, \tag{8.40}$$

where $\mathbf{v} = [v_1 \quad v_2 \quad \cdots \quad v_m]^T$, $\mathbf{H}(\mathbf{v}) = [h_1(v_1) \quad h_2(v_2) \quad \cdots \quad h_m(v_m)]^T$, $\mathbf{H}^{-1}(\mathbf{v}) = [h_1^{-1}(v_1) \quad h_2^{-1}(v_2) \quad \cdots \quad h_m^{-1}(v_m)]^T$, and $h_i(\cdot)$ is a bounded, monotonic, and odd $C^p(p \geqslant 1)$ function $(i = 1, 2, \cdots, m)$. The function of $\mathbf{H}(\mathbf{v})$ is to map the physical limits imposed.

In order to derive the optimal control law, the Hamilton–Jacobi–Bellman (HJB) equation needs to be solved:

$$\nabla V(\mathbf{x})^T[\mathbf{f}(\mathbf{x}) - \mathbf{g}(\mathbf{x})\mathbf{H}(\frac{1}{2}\mathbf{R}^{-1}\mathbf{g}^T(\mathbf{x})\nabla V(\mathbf{x}))] + Q(\mathbf{x}) \\ + 2\int_0^{-\mathbf{H}(\frac{1}{2}\mathbf{R}^{-1}\mathbf{g}^T(\mathbf{x})\nabla V(\mathbf{x}))} (\mathbf{H}^{-1}(\mathbf{v}))^T\mathbf{R}d\mathbf{v} = 0, \tag{8.41}$$

with $V(0) = 0$. If (8.41) has a solution, then

$$u^*(\mathbf{x}) = -\mathbf{H}(\frac{1}{2}\mathbf{R}^{-1}\mathbf{g}^T(\mathbf{x})\nabla V^*(\mathbf{x})). \tag{8.42}$$

However, (8.41) is difficult to solve explicitly. Numerical algorithms, e.g., police iteration (PI) [14], are devised to approximate the solution:

(i) Design an initial admissible controller $u_0(\mathbf{x})$.
(ii) Start from $i \geqslant 0$ and solve the Lyapunov equation to obtain $V(i)$,

$$\nabla V_i(\mathbf{x})^{\mathrm{T}}[\mathbf{f}(\mathbf{x}) + \mathbf{g}(\mathbf{x})u_i(\mathbf{x})] + Q(\mathbf{x}) + 2\int_0^{\mathbf{u}_i(\mathbf{x})} (\mathbf{H}^{-1}(\mathbf{v}))^{\mathrm{T}}\mathbf{R}d\mathbf{v} = 0. \tag{8.43}$$

(iii) Update the control law

$$\mathbf{u}_{i+1}(\mathbf{x}) = -\mathbf{H}(\frac{1}{2}\mathbf{R}^{-1}\mathbf{g}^{\mathrm{T}}(\mathbf{x})\nabla V_i(\mathbf{x})) \tag{8.44}$$

and let $i = i + 1$. Stop the iteration and output $V_i(\mathbf{x})$ and $\mathbf{u}_i$ as the approximated optimal solution and optimal controller when the stopping criterion is met.

Theorem 3. *For system (8.38), starting from the initial controller $\boldsymbol{u}_0(\boldsymbol{x})$, two sequences of $\{V_i(\boldsymbol{x})\}_{i=0}^{\infty}$ and $\{\boldsymbol{u}_{i+1}(\boldsymbol{x})\}_{i=0}^{\infty}$ could be generated via the PI approach defined by (8.43) and (8.44), and*

1. $0 \leqslant V_{i+1}(\boldsymbol{x}) \leqslant V_i(\boldsymbol{x})$,
2. *$\boldsymbol{u}_i(\boldsymbol{x})$ is within the physical limits and admissible, and*
3. *if V^* and $\boldsymbol{u}^*$ exist, then $V_i(\boldsymbol{x}) \to V^*$, $\boldsymbol{u}_i(\boldsymbol{x}) \to \boldsymbol{u}^*$.*

Proof. The proof of this theorem follows the lines of reasoning of **Lemma 1** and **Theorem 1** in [14], and is omitted here for brevity. □

Although the optimal control problem can be solved using the PI, the PI can be applied only when $\mathbf{f}(\mathbf{x})$ and $\mathbf{g}(\mathbf{x})$ are known. However, for some working conditions, the values of model parameters cannot be obtained. As a result, $\mathbf{f}(\mathbf{x})$ and $\mathbf{g}(\mathbf{x})$ are unknown. The optimal control problem is therefore transferred to find the optimal controller without knowing the system dynamics, which relies on the use of model-free PI approaches [15].

8.3.2 Model-free zinc powder dosage controller

In PI, the controller is iteratively improved using (8.43) and (8.44). In model-free PI (MFPI), the controller is iteratively improved by learning from the input-state information. In MFPI, the practical control input $\mathbf{u}(\mathbf{x})$ is decomposed as

$$\mathbf{u}(\mathbf{x}) = \mathbf{u}_i(\mathbf{x}) + \mathbf{v}_i, \tag{8.45}$$

where $\mathbf{u}_i(\mathbf{x})$ is the iterative control policy to be updated and $\mathbf{v}_i$ is the excitation noise. Therefore,

$$\dot{\mathbf{x}} = \mathbf{f}(\mathbf{x}) + \mathbf{g}(\mathbf{x})\mathbf{u}_i(\mathbf{x}) + \mathbf{g}(\mathbf{x})\mathbf{v}_i. \tag{8.46}$$

If $\mathbf{u}_i$ acts on the ACP, then

$$\dot{V}_i = \nabla V_i(\mathbf{x})^T\{[f(\mathbf{x}) + g(\mathbf{x})\mathbf{u}_i(\mathbf{x})] + g(\mathbf{x})\mathbf{v}_i\} \tag{8.47}$$

$$= -Q(\mathbf{x}) - W(\mathbf{u}_i(\mathbf{x})) + \nabla V_i(\mathbf{x})^T g(\mathbf{x})\mathbf{v}_i \tag{8.48}$$

$$= -Q(\mathbf{x}) - W(\mathbf{u}_i(\mathbf{x})) - 2(\mathbf{H}^{-1}(\mathbf{u}_{i+1}(\mathbf{x})))^{\mathrm{T}}\mathbf{R}^{\mathrm{T}}\mathbf{v}_i. \tag{8.49}$$

If we integrate (8.49) on time interval $[t, t+\Delta t]$, then

$$V_i(\mathbf{x}(t+\Delta t)) - V_i(\mathbf{x}(t)) = \int_t^{t+\Delta t} [-Q(\mathbf{x}) - W(\mathbf{u}_i) - S(\mathbf{u}_{i+1})\mathbf{v}_i]d\tau, \tag{8.50}$$

where $S(\mathbf{u}_{i+1}) = 2(\mathbf{H}^{-1}(\mathbf{u}_{i+1}(\mathbf{x})))^{\mathrm{T}}\mathbf{R}^{\mathrm{T}}$.

It can be observed from (8.50) that if $\mathbf{u}_i(\mathbf{x})$ is given, then $V_i(\mathbf{x})$ and $\mathbf{u}_{i+1}(\mathbf{x})$ can be approximated; $\{\phi_j(\mathbf{x})\}_{j=1}^{\infty}$ and $\{\psi_j(\mathbf{x})\}_{j=1}^{\infty}$ are two infinite sequences of linearly independent smooth basis functions on a compact set Ω; $\phi_j(0) = 0$ and $\psi_j(0) = 0$ for all $j = 1, 2, \cdots$; and $V_i(\mathbf{x})$ can be approximated as follows:

$$\hat{V}_i(\mathbf{x}) = \sum_{j=1}^{N_1} \hat{c}_{i,j}\phi_j(\mathbf{x}) = \Phi(\mathbf{x})^{\mathrm{T}}\hat{\mathbf{c}}_i, \tag{8.51}$$

where $\hat{\mathbf{c}}_i = [\hat{c}_{i,1} \quad \hat{c}_{i,2} \quad \cdots \quad \hat{c}_{i,N_1}]^{\mathrm{T}}$ and $\Phi(\mathbf{x}) = [\phi_1(\mathbf{x}) \quad \phi_2(\mathbf{x}) \quad \cdots \quad \phi_{N_1}(\mathbf{x})]^{\mathrm{T}}$. Similarly, $\mathbf{u}_{i+1}(\mathbf{x})$ can be approximated as

$$\hat{\mathbf{u}}_{i+1}(\mathbf{x}) = -\mathbf{H}(\frac{1}{2}\mathbf{R}^{-1}\hat{\Upsilon}_{i+1}(\mathbf{x})), \tag{8.52}$$

where $\mathbf{H}(\cdot)$ is used to guarantee the satisfaction of input constraints,

$$\hat{\Upsilon}_i(\mathbf{x}) = [\upsilon_{i,1}(\mathbf{x}) \quad \upsilon_{i,2}(\mathbf{x}) \quad \cdots \quad \upsilon_{i,m}(\mathbf{x})]^{\mathrm{T}}, \tag{8.53}$$

and for each $\upsilon_{i,j}(\mathbf{x})$, $j = 1, 2, \cdots, m$,

$$\upsilon_{i,j}(\mathbf{x}) = \sum_{l=1}^{N_{2,j}} \hat{k}_{i_{j,l}}\psi_{j,l}(\mathbf{x}) = \Psi_j(\mathbf{x})^{\mathrm{T}}\hat{\mathbf{k}}_{i,j}, \tag{8.54}$$

$$\hat{\mathbf{k}}_{i,j} = [\hat{k}_{i_{j,1}} \quad \hat{k}_{i_{j,2}} \quad \cdots \quad \hat{k}_{i_{j,N_{2,j}}}]^{\mathrm{T}}, \tag{8.55}$$

$$\Psi_j(\mathbf{x}) = [\psi_{j,1}(\mathbf{x}) \quad \psi_{j,2}(\mathbf{x}) \quad \ldots \quad \psi_{j,N_{2,j}}(\mathbf{x})]^{\mathrm{T}}, \tag{8.56}$$

where $N_1 > 0$ and $N_2 = \sum_{j=1}^{m} N_{2,j} > 0$ are sufficiently large integers. Using the approximation and denoting t and $t+\Delta t$ as t_k and t_{k+1}, (8.50) can be reformulated as

$$\sum_{j=1}^{N_1} \hat{c}_{i,j}[\phi_j(\mathbf{x}(t_{k+1})) - \phi_j(\mathbf{x}(t_k))] - \int_{t_k}^{t_{k+1}} \sum_{j=1}^{m} \hat{\mathbf{k}}_{i+1,j}^{\mathrm{T}}\Psi_j(\mathbf{x})(u_j(\mathbf{x}) - \hat{u}_j(\mathbf{x}))dt$$

$$= -\int_{t_k}^{t_{k+1}} [Q(\mathbf{x}) + W(\hat{\mathbf{u}}_i(\mathbf{x}))]dt + \varepsilon_{i,k}. \tag{8.57}$$

Considering the time sequence $\{t_k\}_{k=0}^{L}$ with $L \geqslant N_1 + N_2$, we have

$$[\Xi_\Phi \quad -\mathbf{I}_{\Psi\mathbf{u}} + \mathbf{I}_{\Psi\hat{\mathbf{u}}_i}] \begin{bmatrix} \hat{\mathbf{c}}_i \\ \hat{\mathbf{k}}_{i+1,1} \\ \hat{\mathbf{k}}_{i+1,2} \\ \vdots \\ \hat{\mathbf{k}}_{i+1,m} \end{bmatrix} = -\mathbf{M}_Q - \mathbf{M}_W + \Delta_i, \tag{8.58}$$

where

$$\Xi_\Phi = \begin{bmatrix} \Phi(\mathbf{x}(1))^{\mathrm{T}} - \Phi(\mathbf{x}(0))^{\mathrm{T}} \\ \Phi(\mathbf{x}(2))^{\mathrm{T}} - \Phi(\mathbf{x}(1))^{\mathrm{T}} \\ \vdots \\ \Phi(\mathbf{x}(L))^{\mathrm{T}} - \Phi(\mathbf{x}(L-1))^{\mathrm{T}} \end{bmatrix}_{L\times N_1}, \tag{8.59}$$

$$\mathbf{I}_{\Psi\mathbf{u}} = \begin{bmatrix} \int_{t_0}^{t_1} [\Psi_1(\mathbf{x})^{\mathrm{T}} u_1(\mathbf{x}) \quad \Psi_2(\mathbf{x})^{\mathrm{T}} u_2(\mathbf{x}) \cdots \Psi_m(\mathbf{x})^{\mathrm{T}} u_m(\mathbf{x})] dt \\ \int_{t_1}^{t_2} [\Psi_1(\mathbf{x})^{\mathrm{T}} u_1(\mathbf{x}) \quad \Psi_2(\mathbf{x})^{\mathrm{T}} u_2(\mathbf{x}) \cdots \Psi_m(\mathbf{x})^{\mathrm{T}} u_m(\mathbf{x})] dt \\ \vdots \\ \int_{t_{L-1}}^{t_L} [\Psi_1(\mathbf{x})^{\mathrm{T}} u_1(\mathbf{x}) \quad \Psi_2(\mathbf{x})^{\mathrm{T}} u_2(\mathbf{x}) \cdots \Psi_m(\mathbf{x})^{\mathrm{T}} u_m(\mathbf{x})] dt \end{bmatrix}_{L\times N_2}, \tag{8.60}$$

$$\mathbf{I}_{\Psi\hat{\mathbf{u}}_i} = \begin{bmatrix} \int_{t_0}^{t_1} [\Psi_1(\mathbf{x})^{\mathrm{T}} \hat{u}_1(\mathbf{x}) \quad \Psi_2(\mathbf{x})^{\mathrm{T}} \hat{u}_2(\mathbf{x}) \cdots \Psi_m(\mathbf{x})^{\mathrm{T}} \hat{u}_m(\mathbf{x})] dt \\ \int_{t_1}^{t_2} [\Psi_1(\mathbf{x})^{\mathrm{T}} \hat{u}_1(\mathbf{x}) \quad \Psi_2(\mathbf{x})^{\mathrm{T}} \hat{u}_2(\mathbf{x}) \cdots \Psi_m(\mathbf{x})^{\mathrm{T}} \hat{u}_m(\mathbf{x})] dt \\ \vdots \\ \int_{t_{L-1}}^{t_L} [\Psi_1(\mathbf{x})^{\mathrm{T}} \hat{u}_1(\mathbf{x}) \quad \Psi_2(\mathbf{x})^{\mathrm{T}} \hat{u}_2(\mathbf{x}) \cdots \Psi_m(\mathbf{x})^{\mathrm{T}} \hat{u}_m(\mathbf{x})] dt \end{bmatrix}_{L\times N_2}, \tag{8.61}$$

$$\mathbf{M}_Q = \begin{bmatrix} \int_{t_0}^{t_1} Q(\mathbf{x}) dt \\ \int_{t_1}^{t_2} Q(\mathbf{x}) dt \\ \vdots \\ \int_{t_{L-1}}^{t_L} Q(\mathbf{x}) dt \end{bmatrix}_{L\times 1}, \tag{8.62}$$

$$\mathbf{M}_W = \begin{bmatrix} \int_{t_0}^{t_1} W(\hat{\mathbf{u}}_i)dt \\ \int_{t_1}^{t_2} W(\hat{\mathbf{u}}_i)dt \\ \vdots \\ \int_{t_{L-1}}^{t_L} W(\hat{\mathbf{u}}_i)dt \end{bmatrix}_{L\times 1}, \tag{8.63}$$

and $\Delta_i = [\varepsilon_{i,1} \quad \varepsilon_{i,1} \quad \dots \quad \varepsilon_{i,L}]^{\mathrm{T}} \in \mathbb{R}^L$ is the approximation error.

Therefore, the approximation problem is transferred to the estimation of $\hat{\mathbf{c}}_i$ and $\hat{\mathbf{k}}_{i+1}$. If

$$\operatorname{rank}(\Theta_i) = N_1 + N_2, \tag{8.64}$$

then $\hat{\mathbf{c}}_i$ and $\hat{\mathbf{k}}_{i+1}$ can be estimated using the least square

$$\begin{bmatrix} \hat{\mathbf{c}}_i \\ \hat{\mathbf{k}}_{i+1} \end{bmatrix} = (\Theta^{\mathrm{T}}\Theta)^{-1}\Theta^{\mathrm{T}}\Pi, \tag{8.65}$$

in which $\Theta = [\Xi_\Phi \quad -\mathbf{I}_{\Psi u} + \mathbf{I}_{\Psi \hat{u}_i}]$, $\Pi = [-\mathbf{M}_Q - \mathbf{M}_W]$, and $\hat{\mathbf{k}}_{i+1} = [\hat{\mathbf{k}}_{i,1}^{\mathrm{T}} \quad \hat{\mathbf{k}}_{i,2}^{\mathrm{T}} \quad \cdots \quad \hat{\mathbf{k}}_{i,m}^{\mathrm{T}}]^{\mathrm{T}}$.

Note that $\mathbf{I}_{\Psi \hat{\mathbf{u}}_i}$ and $\mathbf{M}_W$ are functions of $\hat{\mathbf{k}}_i$. Therefore, if $[\mathbf{x}, \mathbf{u}]$ on the learning interval $[t_0, t_l]$ is obtained, then PI can be conducted offline.

Lemma 1. *If* **Assumption 1** *and* **Assumption 2** *hold, then for each i and arbitrary $\epsilon > 0$, there exist integers $N_1^{**} > 0$ and $N_2^{**} > 0$:*

$$\left| \sum_{j=1}^{N_1} \hat{c}_{i,j}\phi_j(\boldsymbol{x}) - V_i(\boldsymbol{x}) \right| \leqslant \epsilon, \tag{8.66}$$

$$\left| -\boldsymbol{H}(\frac{1}{2}\boldsymbol{R}^{-1}\hat{\Upsilon}_{i+1}(\boldsymbol{x})) - \boldsymbol{u}_{i+1}(\boldsymbol{x}) \right| \leqslant \epsilon, \tag{8.67}$$

for all $x \in \Omega$, if $N_1 > N_1^{**}$ and $N_2 > N_2^{**}$.

Proof. The proof of this lemma follows the same lines of reasoning of **Theorem 3.1** in [15], and is omitted here for brevity. □

Theorem 4. *If* **Assumption 1** *and* **Assumption 2** *hold, then by applying the MFPI, the sequences of $\{\hat{V}_i(\boldsymbol{x})\}_{i=0}^{\infty}$ and $\{\hat{\boldsymbol{u}}_{i+1}(\boldsymbol{x})\}_{i=0}^{\infty}$ can be generated; $\hat{\boldsymbol{u}}_i(\boldsymbol{x})$ is admissible and satisfies the input constraints. In addition, if $V^*(\boldsymbol{x})$ and $\boldsymbol{u}^*(\boldsymbol{x})$ exist, then for arbitrary $\rho > 0$, there exist $i^* > 0$, $N_1^* > 0$, and $N_2^* > 0$ such that if $N_1 > N_1^*$ and $N_2 > N_2^*$,*

$$|\hat{V}_i^*(\boldsymbol{x}) - V^*(\boldsymbol{x})| \leqslant \rho, \tag{8.68}$$

$$|\hat{\boldsymbol{u}}_{i^*+1}(\boldsymbol{x}) - \boldsymbol{u}^*(\boldsymbol{x})| \leqslant \rho \tag{8.69}$$

for all $\boldsymbol{x} \in \Omega$.

Proof. According to **Theorem 1**, there exists $i^* > 0$ which satisfies

$$|V_{i^*}(\mathbf{x}) - V^*(\mathbf{x})| \leqslant \frac{\rho}{2}, \tag{8.70}$$

$$|\mathbf{u}_{i^*+1}(\mathbf{x}) - \mathbf{u}^*(\mathbf{x})| \leqslant \frac{\rho}{2} \tag{8.71}$$

for any $\mathbf{x} \in \Omega$. According to **Lemma 1**, there exist integers $N_1^* > 0$ and $N_2^* > 0$, for any $N_1 > N_1^*$ and $N_2 > N_2^*$,

$$\left|\sum_{j=1}^{N_1} \hat{c}_{i^*,j}\phi_j(\mathbf{x}) - V_{i^*}(\mathbf{x})\right| \leqslant \frac{\rho}{2}, \tag{8.72}$$

$$\left|-\mathbf{H}(\frac{1}{2}\mathbf{R}^{-1}\hat{\Upsilon}_{i^*+1}(\mathbf{x})) - \mathbf{u}_{i^*+1}(\mathbf{x})\right| \leqslant \frac{\rho}{2}. \tag{8.73}$$

Therefore, according to the triangle inequality,

$$|\hat{V}_i^*(\mathbf{x}) - V^*(\mathbf{x})| \leqslant \rho,$$

$$|\hat{\mathbf{u}}_{i^*+1}(\mathbf{x}) - \mathbf{u}^*(\mathbf{x})| \leqslant \rho.$$

Assumption 1: There exist $L_0 > 0$ and $\delta > 0$ such that for all $L > L_0$,

$$\frac{1}{L}\sum_{k=1}^{l} \theta_{i,k}^{\mathrm{T}}\theta_{i,k} \geqslant \mathrm{I}_{\mathrm{N_1+N2}}, \tag{8.74}$$

where

$$\theta_{i,k}^{\mathrm{T}} = \begin{bmatrix} \Phi(\mathbf{x}(k+1)) - \Phi(\mathbf{x}(k)) \\ \int_{t_k}^{t_{k+1}} [\Psi_1(\mathbf{x})^{\mathrm{T}}\hat{v}_1(\mathbf{x}) \quad \Psi_2(\mathbf{x})^{\mathrm{T}}\hat{v}_2(\mathbf{x}) \cdots \Psi_m(\mathbf{x})^{\mathrm{T}}\hat{v}_m(\mathbf{x})]^{\mathrm{T}} dt \end{bmatrix}. \tag{8.75}$$

Assumption 2: The closed-loop ACP is ISS when the exploration noise is considered as the input [12].

As shown in Fig. 8.9, the MFPI algorithm is composed of three stages:

(i) Initialization stage: Design the initial admissible controller, select the basis functions, and determine the design parameters.

(ii) Online learning stage: Apply the initial admissible control law on ACP and collect the state and input information pair during the learning stage. When the full column condition is met, stop the learning stage.

(iii) Offline policy iteration stage: Iteratively update the control law by learning from the collected information until the stop criterion is met.

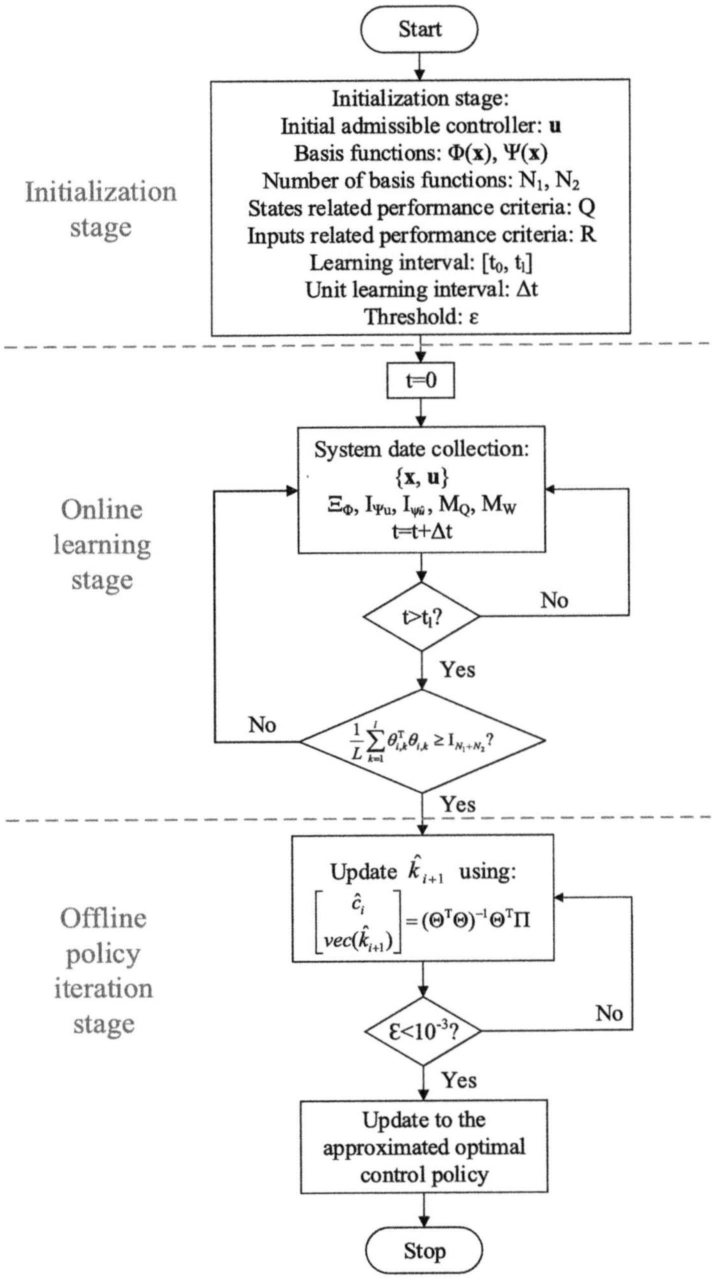

FIGURE 8.9 Flowchart of the ADP algorithm.

8.3.3 Control performance analysis

In order to test the control performance of the proposed approach, data samples collected from the site were utilized to conduct an experimental study. For confidential reasons, the data are scaled and desensitized. The ACP is simulated using a model. Four working conditions with different model parameters are included in the study. The control targets under the four different working conditions are $[6.0660, 0.9577, 0.4522, 0.2500]$, $[4.0500, 0.7225, 0.3850, 0.2500]$, $[4.800, 0.8100, 0.4100, 0.2500]$, and $[6.2070, 0.9742, 0.4569, 0.2500]$.

In the simulation, the control algorithm is first configured in the initialization stage. The optimal control objective is to minimize

$$J = \int_0^{\infty} (\mathbf{x}^{\mathrm{T}}\mathbf{Q}\mathbf{x} + \mathbf{u}^{\mathrm{T}}\mathbf{R}\mathbf{u})d\tau, \tag{8.76}$$

where $\mathbf{Q} \in \mathbb{R}^{8\times 8} = \mathrm{diag}(1, 0, 1, 0, 1, 0, 1, 0)$ and $\mathbf{R} = \mathrm{diag}(0.5, 0.5, 0.5, 0.5)$. The initial value of $\mathbf{x}$ was $[1.50, -0.530, 0.50, -0.570, 0.25, -0.630, 0.15, -0.670]^{\mathrm{T}}$. The length of the learning interval was chosen as 0.5 hours. The sampling rate of process variables was every 30 seconds. The stopping criterion is that the norm of $\hat{\mathbf{k}}$ is less than $\epsilon = 10^{-3}$.

The initial admissible control law was

$$u_i = \Gamma(i) + \Lambda(i)\tanh\left(\frac{-2\mathrm{x}_{2\mathrm{i}-1} - \mathrm{x}_{2\mathrm{i}-1}^2}{\Lambda(\mathrm{i})} + \mathrm{e}(\mathrm{i})\right), \tag{8.77}$$

where $\Gamma(i)$ and $\Lambda(i)$ were decided according to operational experience. The exploration noise was

$$e = [\sin(0.01\mathrm{t}) \quad \sin(0.01\mathrm{t}) \quad \sin(0.01\mathrm{t}) \quad \sin(0.01\mathrm{t})]^{\mathrm{T}} \tag{8.78}$$

under status 1 and

$$e = [\sin(0.02\mathrm{t}) \quad \sin(0.02\mathrm{t}) \quad \sin(0.02\mathrm{t}) \quad \sin(0.02\mathrm{t})]^{\mathrm{T}} \tag{8.79}$$

under status 2–4.

The basis functions include

$$\Phi(\mathbf{x}) = [x_{2i-1}^2 \quad x_{2i}^2 \quad x_{2i-1}x_{2i} \quad x_{2i-1}^4 \quad x_{2i}^4 \quad x_{2i-1}^3 x_{2i} x_{2i-1}^2 x_{2i}^2 \quad x_{2i-1}x_{2i}^3]^{\mathrm{T}}, \tag{8.80}$$

$$\Psi_i(\mathbf{x}) = [x_{2i-1} \quad x_{2i-1}^2]^{\mathrm{T}}. \tag{8.81}$$

In the learning stage, the admissible controller is forced on the system. After sufficient "input-state" information is collected in the learning state, an approximated optimal control law is obtained by the MFPI in the offline policy iteration stage. As shown in Figs. 8.10–8.13, compared with the initial admissi-

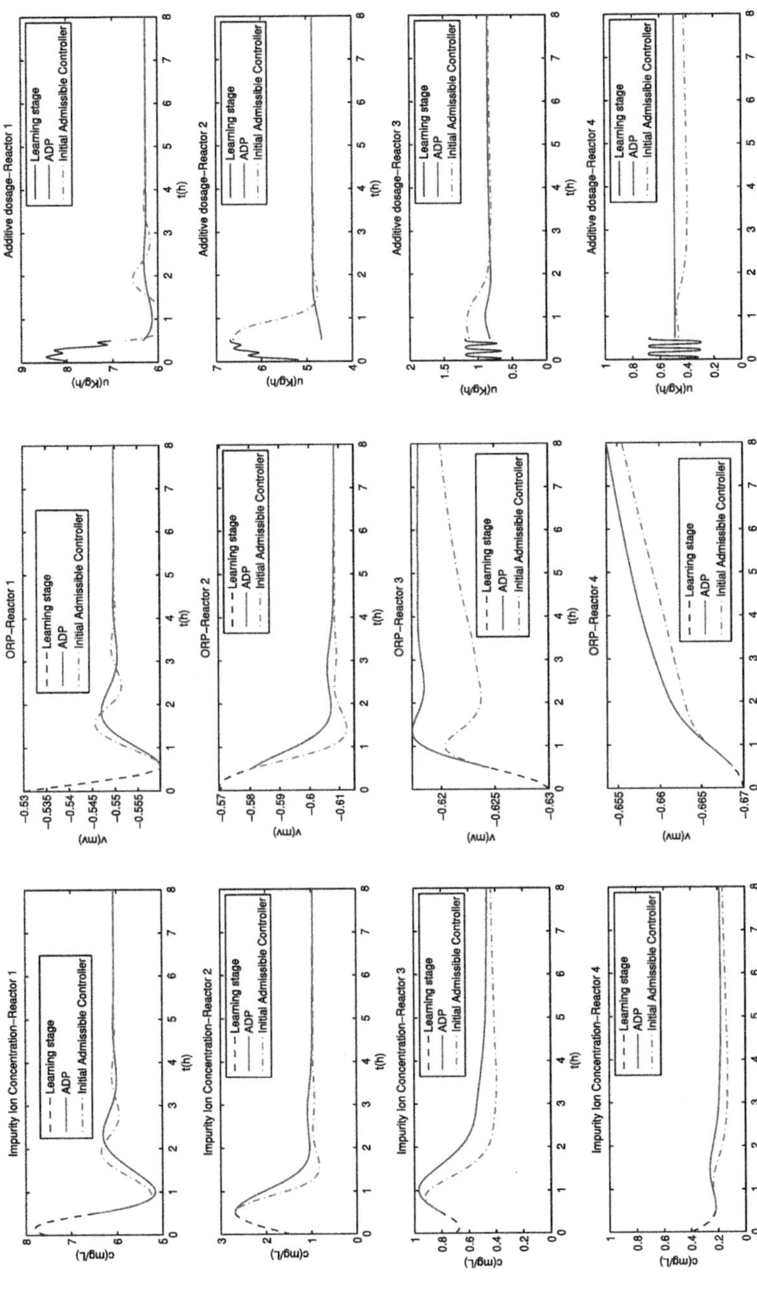

FIGURE 8.10 Control performance under working condition 1.

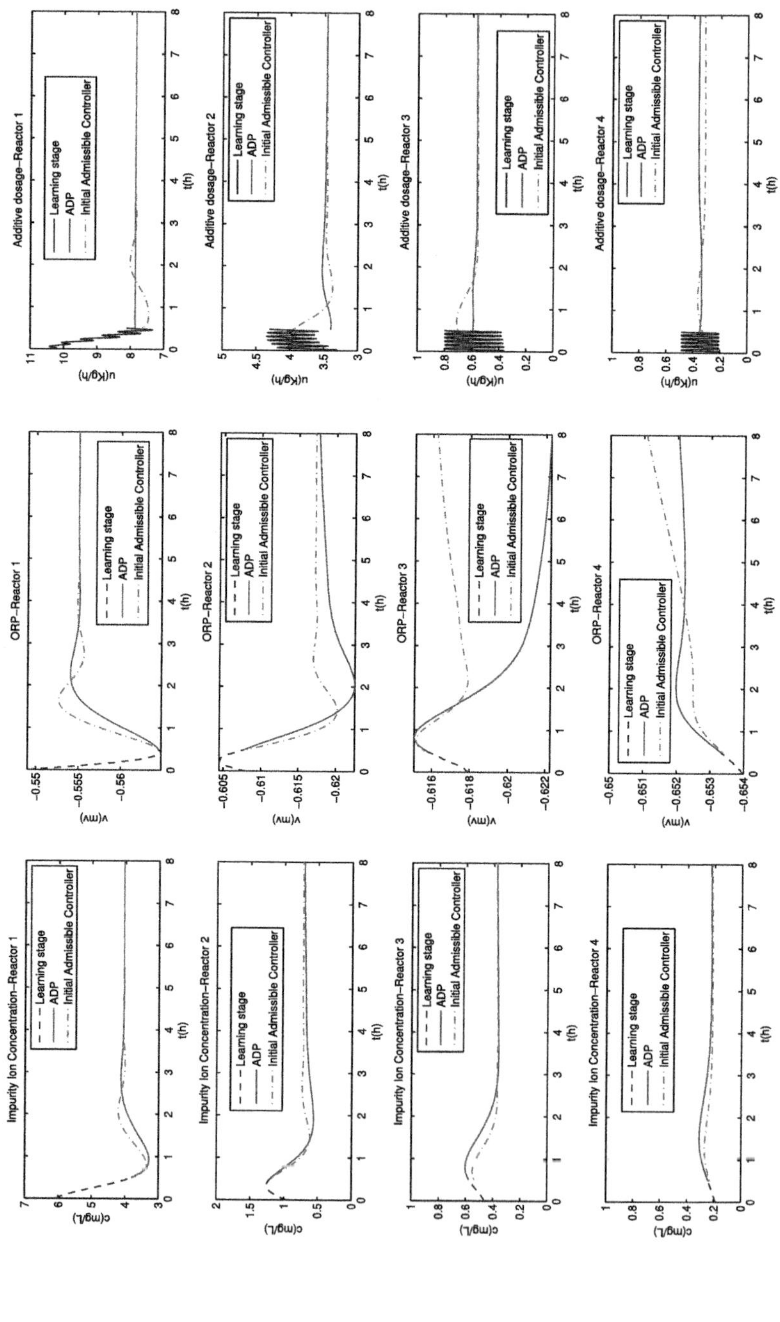

FIGURE 8.11 Control performance under working condition 2.

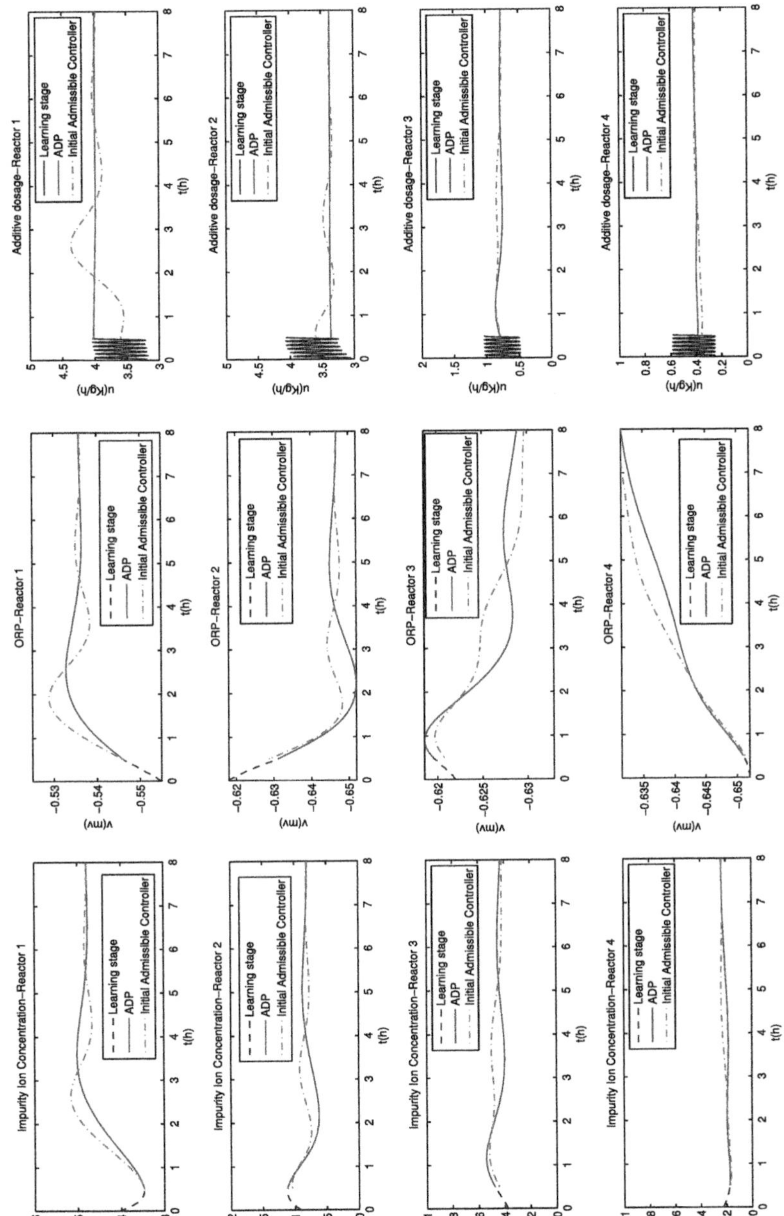

FIGURE 8.12 Control performance under working condition 3.

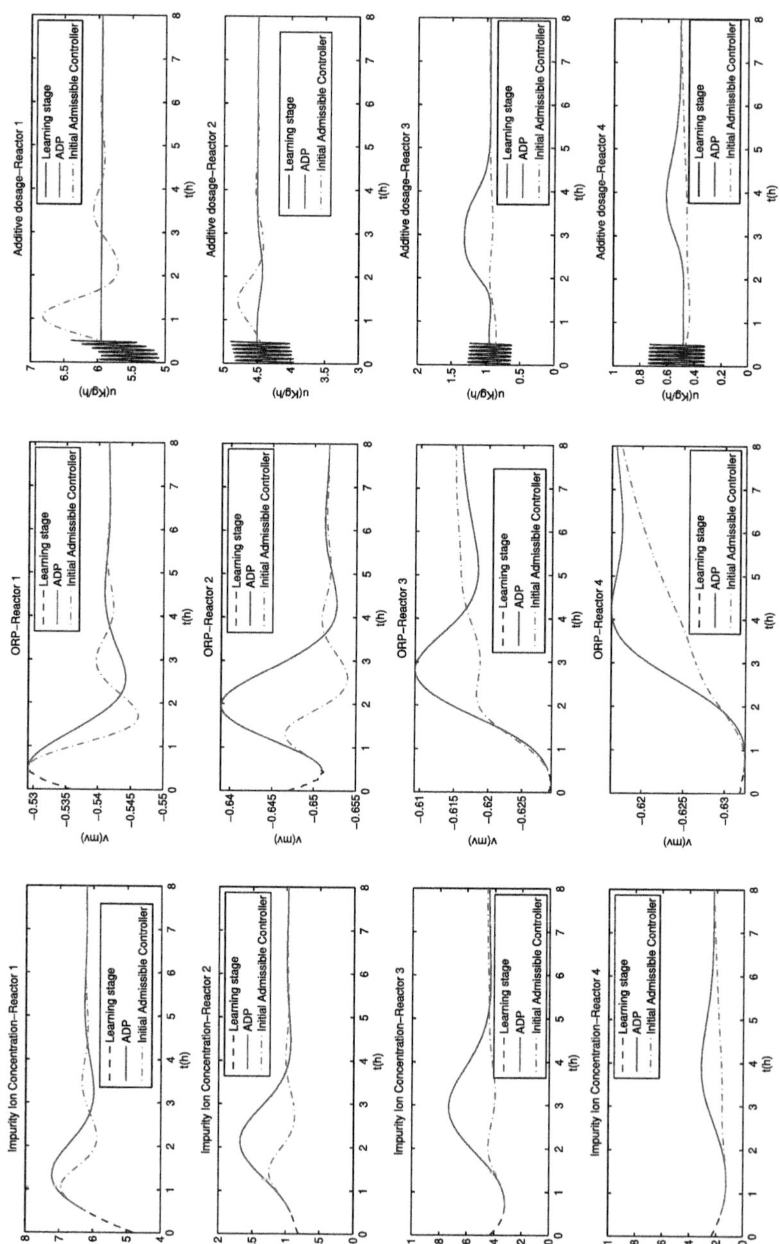

FIGURE 8.13 Control performance under working condition 4.

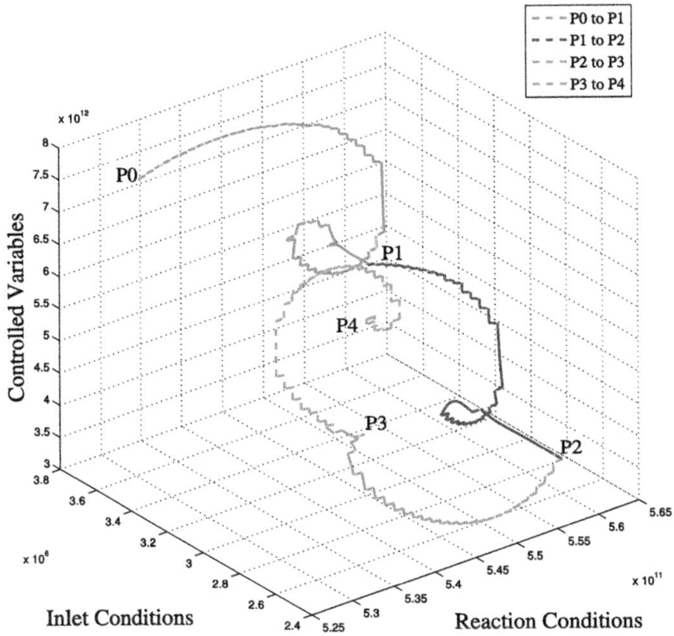

FIGURE 8.14 Running trajectory of the ACP in the comprehensive state space.

ble control law, state trajectories using ADP are more smooth. In addition, the system states have less fluctuations. The resulting steady-state outlet cobalt ion concentrations under the four working conditions are $[6.066, 0.9607, 0.4615, 0.187]$, $[4.049, 0.7073, 0.3775, 0.226]$, $[4.790, 0.8036, 0.4303, 0.2351]$, and $[6.201, 0.9719, 0.4423, 0.224]$. The absolute relative errors between the resulting state and their targets are $[0.01\%, 0.31\%, 2.05\%, 25.16\%]$, $[0, 2.10\%, 1.95\%, 9.68\%]$, $[0.21\%, 0.79\%, 4.96\%, 5.95\%]$, and $[0.009\%, 0.23\%, 3.18\%, 10.22\%]$. This indicates the convergency and a good tracking performance. To illustrate the movement of ACP during the test, the trajectory of the process is mapped into the "comprehensive state space" (Fig. 8.14). In the comprehensive state space, the output states are outlet cobalt ion concentrations of each reactor. The reaction conditions, which include various process variables, are simply expressed using the ORPs of each reactor. The inlet conditions include the flow rate and cobalt ion concentration of the inlet solution. The resulting control laws are listed in Table 8.1. It is noted that the zinc powder dosages of each reactor are within the practical operating limitations. To sum up, by using the model-free and input-constrained control design approach, an approximated optimal control law can be determined without knowing the system dynamics. Due to its "model-free" feature, it can be used as an alternative when new working conditions emerge or in the pilot-scale testing stage.

TABLE 8.1 The controller lookup table obtained in the simulation study.

Case	Additive dosage controller
P0 to P1	$\mathbf{u}=\begin{bmatrix} V[13.93+5.570\tanh(\frac{0.360996x_1+0.011805x_1^2}{5.570})] \\ V[10.83+4.331\tanh(\frac{-0.192069x_1-0.028850x_1^2}{4.331})] \\ V[1.90+0.760\tanh(\frac{-0.938720x_1+2.125868x_1^2}{0.760})] \\ V[1.09+0.437\tanh(\frac{-0.018080x_1+0.918263x_1^2}{0.437})] \end{bmatrix}$
P1 to P2	$\mathbf{u}=\begin{bmatrix} V[17.46+6.986\tanh(\frac{0.028445_1+0.048694_1^2}{6.986})] \\ V[7.68+3.072\tanh(\frac{-0.768491_1+0.993532_1^2}{3.072})] \\ V[1.27+0.508\tanh(\frac{0.649973_1-2.209625_1^2}{0.508})] \\ V[0.771+0.308\tanh(\frac{-0.974369_1+10.056377_1^2}{0.308})] \end{bmatrix}$
P2 to P3	$\mathbf{u}=\begin{bmatrix} V[8.87+3.549\tanh(\frac{-0.073665x_1-0.020191_1^2}{3.549}]) \\ V[7.49+2.997\tanh(\frac{-0.050698_1+0.013196_1^2}{2.997})] \\ V[1.68+0.674\tanh(\frac{1.002691x_1+6.614536x_1^2}{0.674})] \\ V[0.94+0.375\tanh(\frac{0.608990x_1-4.833444x_1^2}{0.375})] \end{bmatrix}$
P3 to P4	$\mathbf{u}=\begin{bmatrix} V[13.22+5.287\tanh(\frac{-0.000904x_1+0.010403x_1^2}{5.287})] \\ V[10.00+4.001\tanh(\frac{0.014068_1-0.363393_1^2}{4.001})] \\ V[2.13+0.853\tanh(\frac{1.753754x_1+11.594710x_1^2}{0.853})] \\ V[1.18+0.473\tanh(\frac{2.260971x_1+10.592536x_1^2}{0.473})] \end{bmatrix}$

References

[1] B. Boyanov, V. Konareva, N. Kolev, Removal of cobalt and nickel from zinc sulphate solutions using activated cementation, Journal of Mining and Metallurgy, Section B: Metallurgy 40 (1) (2004) 41–55.

[2] O. Bøckman, T. Østovld, Products formed during cobalt cementation on zinc in zinc sulphate electrolytes, Hydrometallurgy 54 (2000) 65–78.

[3] S. Fugleberg, E. Yllö, A. Jarvinen, Recent development in solution purification at Outokumpu zinc plant, Kokkola, World Zinc 93 (1993) 241–247.

[4] A. Nelson, W. Wang, G.P. Demopoulos, G. Houlachi, The removal of cobalt from zinc electrolyte by cementation: a critical review, Mineral Processing and Extrative Metallurgy Review 20 (1) (2000) 325–356.

[5] A.M. Polcaro, S. Palmas, S. Dernini, Kinetics of cobalt cementation on zinc powder, Industrial & Engineering Chemistry Research 34 (9) (1995) 3090–3095.

[6] B. Sun, W. Gui, T. Wu, Y. Wang, C. Yang, An integrated prediction model of cobalt ion concentration based on oxidation reduction potential, Hydrometallurgy 140 (2013) 102–110.

[7] K. Tozawa, T. Nishimura, M. Akahori, Miguel A. Malaga, Comparison between purification processes for zinc leach solutions with arsenic and antimony trioxides, Hydrometallurgy 30 (1) (1992) 445–461.
[8] B. Sun, W. Gui, Y. Wang, C. Yang, M. He, A gradient optimization scheme for solution purification process, Control Engineering Practice 44 (2015) 89–103.
[9] B. Sun, M. He, Y. Wang, W. Gui, C. Yang, Q. Zhu, A data-driven optimal control approach for solution purification process, Journal of Process Control 68 (2018) 171–185.
[10] C. Wen, J. Zhou, Z. Liu, H. Su, Robust adaptive control of uncertain nonlinear systems in the presence of input saturation and external disturbance, IEEE Transactions on Automatic Control 56 (7) (2011) 1672–1678.
[11] A. Rincón, C. Erazo, F. Angulo, A robust adaptive controller for an anaerobic digester with saturated input: guarantees for the boundedness and convergence properties, Journal of Process Control 22 (9) (2012) 1785–1792.
[12] H.K. Khalil, Nonlinear Systems, 3rd edition, Prentice Hall, New Jersey, 2002, pp. 174–180.
[13] S.E. Lyashevskiy, Constrained optimization and control of nonlinear systems: new results in optimal control, in: Proceedings of the 35th IEEE Conference on Decision and Control, vol. 1, 1996, IEEE, 1996, pp. 541–546.
[14] M. Abu-Khalaf, F.L. Lewis, Nearly optimal control laws for nonlinear systems with saturating actuators using a neural network HJB approach, Automatica 41 (5) (2005) 779–791.
[15] Y. Jiang, Z.P. Jiang, Robust adaptive dynamic programming and feedback stabilization of nonlinear systems, IEEE Transactions on Neural Networks and Learning Systems 25 (5) (2014) 882–893.

Part IV

System development and future research

Chapter 9

Intelligent control system development

Contents

9.1 Framework of intelligent control systems

In order to realize monitoring and control of zinc hydrometallurgical purification processes, an intelligent control system software (ICS) is developed and integrated with the distributed control system (DCS), which forms the control system of the process. As shown in Fig. 9.1, the overall control system consists of three levels, including the sensor and actuator level, the basic automation level, and the monitoring and control level. The sensor and actuator level is composed of sensors which measure flow rates, temperatures, and ORPs and actuators which execute the control command, e.g., valves and belt weighters. The basic automation level is composed of PLC blocks, which sends the sensor signals to the server and action signals to the actuators according to the setting values of manipulated variables. The monitoring and control level is composed of server stations, engineering stations, and operator stations. The server station stores the measured values of process variables. The production data are also sent to the data center of the plant for storage and management purposes. An OLE for process control (OPC) server is installed in the data center. Therefore, production data can be sent to other systems via OPC communication. The engineering station configures the PLC blocks and graphic interface of the configuration software. The operator station provides a platform for the operators to monitor and control the process. However, due to a lack of optimal control algorithms, the DCS system cannot provide optimized setting values of the manipulated variables. The feasibility of the setting values of manipulated variables relies on the experience of human operators.

The ICS is located on the same level with the operator stations of the DCS system. The framework of ICS is shown in Fig. 9.2. The class diagram of the CMainFrame class is shown in Fig. 9.3. It mainly includes four

Modeling, Optimization, and Control of Zinc Hydrometallurgical Purification Process
https://doi.org/10.1016/B978-0-12-819592-5.00021-1

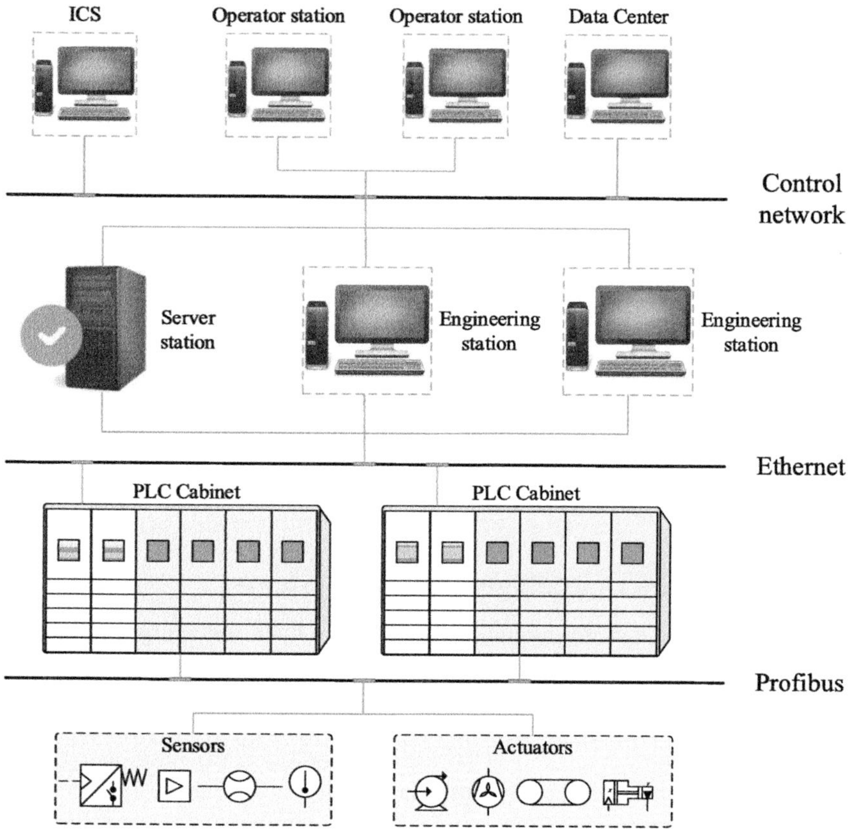

FIGURE 9.1 Structure of the control system.

parts, including data acquisition, data management, process monitoring, and control variables setting. It obtains real-time values of process variables collected by sensors and the chemical test results, which are recorded by the operators or the laboratory information management system via OPC communication. The ICS analyzes/organizes the production data and presents the analysis results on its front-end interfaces, e.g., prediction values of outlet impurity ion concentrations. It runs the optimal control approaches proposed in this book using background programs and provides optimized setting values of the manipulated variables. The core function of ICS is to provide setting values of manipulated variables for the purification process, e.g., ORP/zinc powder dosage, flow rate of arsenic trioxide and spent acid, etc. In addition, to facilitate the operation and meet the site requirement, the ICS also provides functions like data management, process monitoring, fault alarm, etc. Therefore, the operators can handle faults in time and be aware of the production situation.

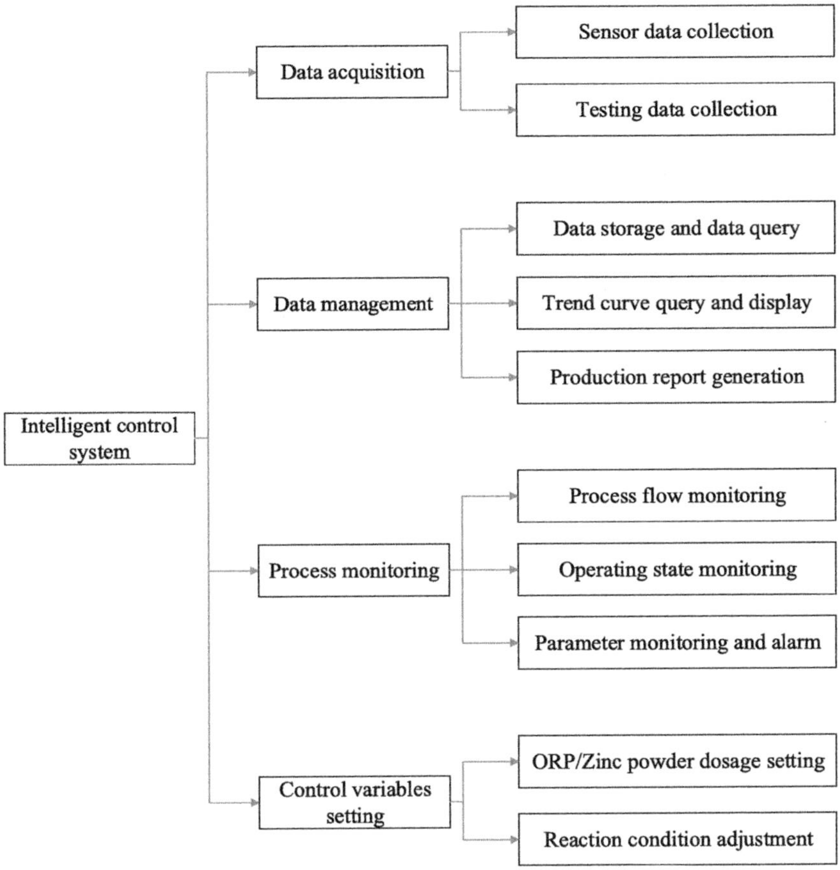

FIGURE 9.2 Functional structure of the ICS.

9.2 Data acquisition and management

The data acquisition part is the foundation of ICS. It includes sensor data collection and test data collection. The sensor data and test data are collected using the OPC client, e.g., solution temperature, pH, ORP, zinc powder dosage, flow rate of feeding solution, etc. These real-time values of process variables are important for process monitoring and setting of manipulated variables. The interface of the OPC client is shown in Fig. 9.4. The ICS also provides an interface for operators to record chemical test results.

The data management part manages the data collected by the data acquisition part. It includes a data storage and data query module, a trend curve query and display module, and a production report generation module. The data storage and data query module stores the collected data into 11 different tables of the database, and provides a query interface for key process variables. The trend

FIGURE 9.3 Class diagram of the CMainFrame class.

curve query and display module can search and illustrate the intuitive variation trajectory of selected variables in the interval of interest. The production report generation module can generate the operation record of the current shift. It can also generate reports of ORPs, zinc powder dosages, flow rates, and chemical

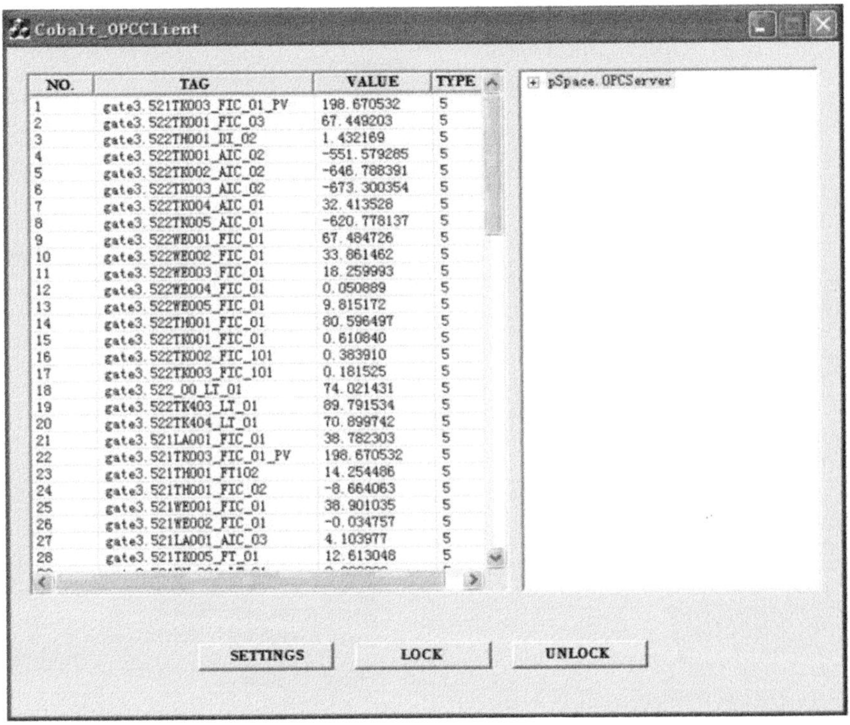

FIGURE 9.4 OPC client.

test results within intervals of interest. The class diagram of the CPrintReportDlg class is shown in Fig. 9.5.

9.3 Process monitoring and control

The control variables setting part includes an ORP/zinc powder dosage setting module and a reaction condition adjustment module. The ORP/zinc powder dosage setting module calculates the optimal setting value of the ORP/zinc powder dosage of each reactor. The reaction condition adjustment module provides reasonable setting values of process variables determining reaction conditions, i.e., flow rates of arsenic trioxide, spent acid, etc. These functions are realized using a background thread which runs simultaneously with front-end threads.

The process monitoring part monitors the key process variables and the operating state. It mainly includes a process flow monitoring module, an operating state monitoring module, and a parameter monitoring and alarm module. The process flow monitoring module is the main interface of the ICS. It displays main technical indices, manipulated variables, and status of equipment. For example, in Figs. 9.6 and 9.7, the setting values of ORPs of each reactor, flow rates of spent acid, and underflow are presented on the interface with label OV

FIGURE 9.5 Class diagram of the CPrintReportDlg class.

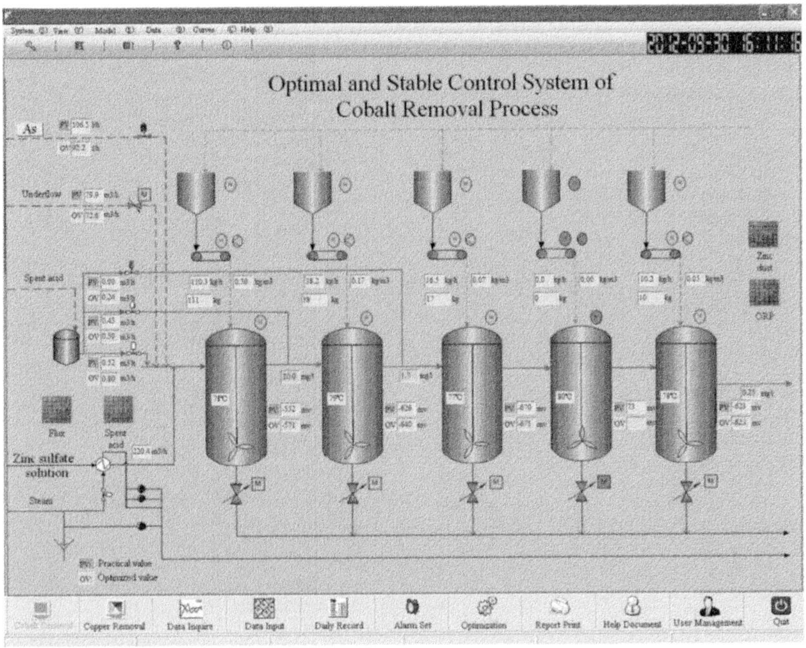

FIGURE 9.6 Process flow monitoring interface of the cobalt removal process.

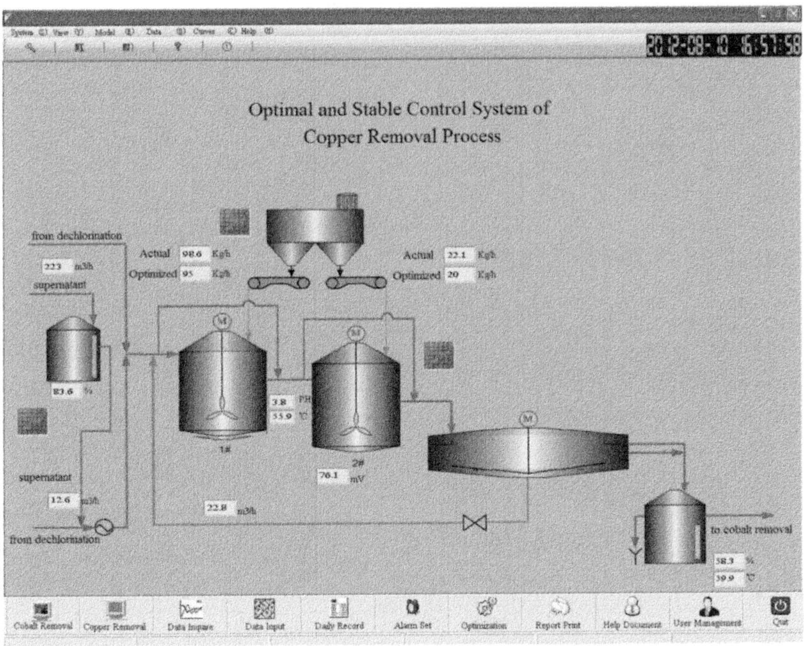

FIGURE 9.7 Process flow monitoring interface of the copper removal process.

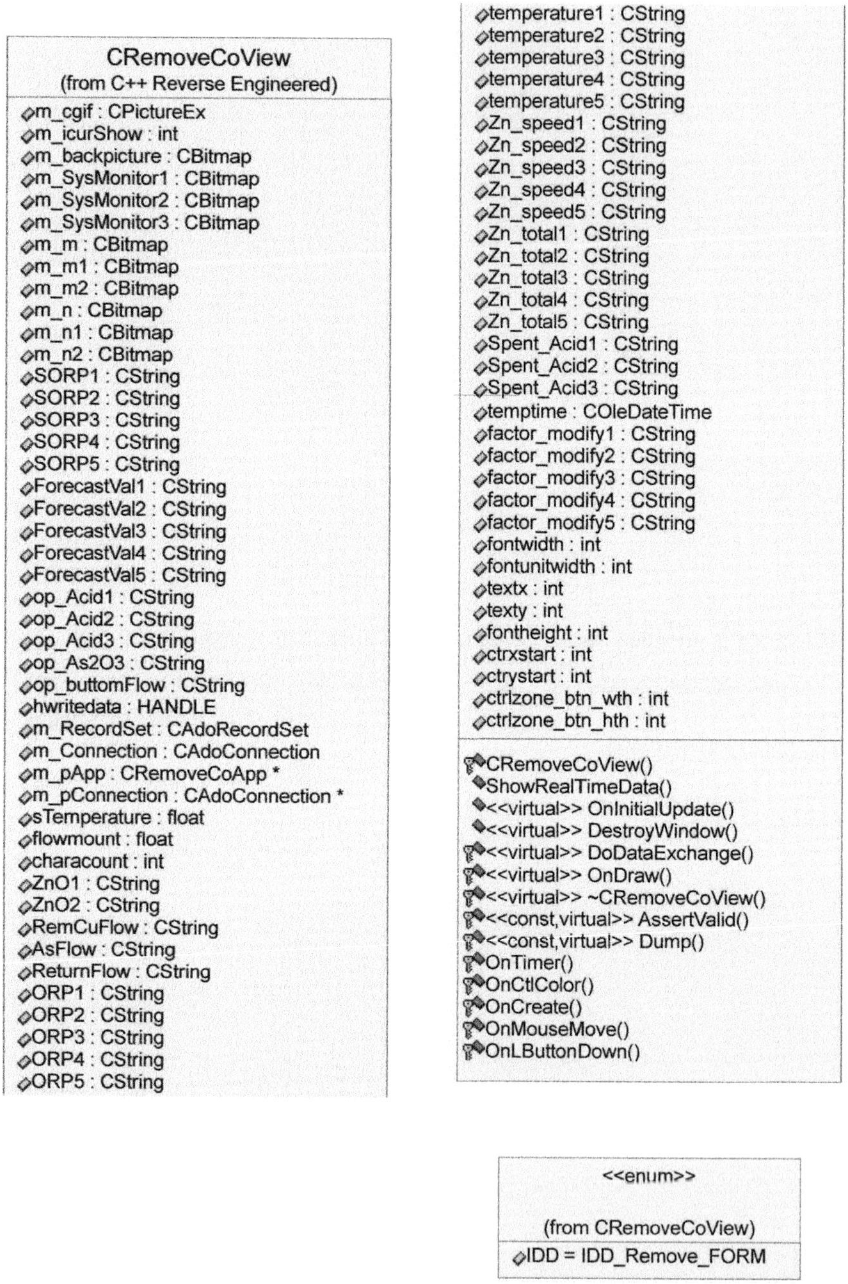

FIGURE 9.8 Class diagram of the CRemoveCoView class.

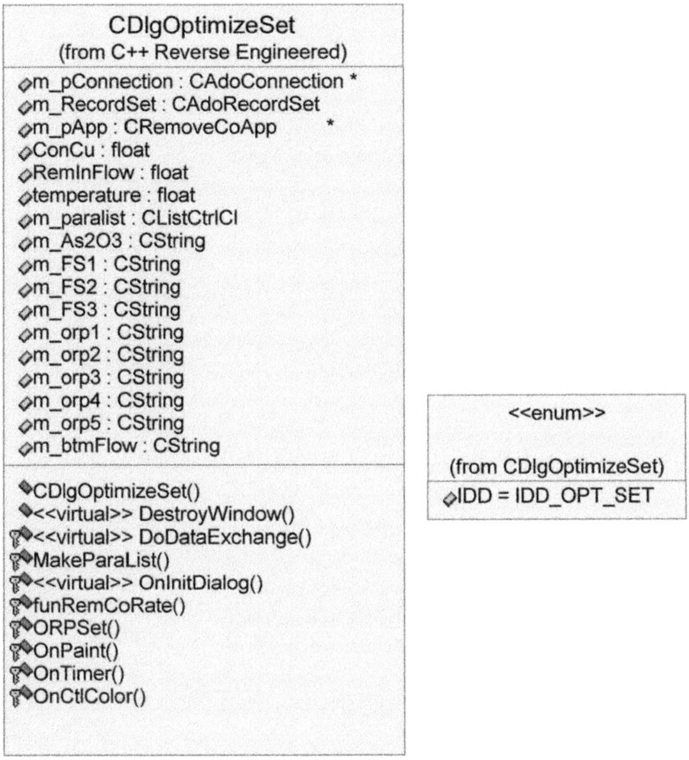

FIGURE 9.9 Class diagram of the CDlgOptimizeSet class.

(optimal value), while PV indicates the practical value. The estimated values of outlet cobalt ion concentrations of #1, #2, and the last reactor are also displayed in Fig. 9.6. The class diagrams of classes CRemoveCoView (for monitoring the cobalt removal process) and CDlgOptimizeSet (for optimization of the ORP) are shown in Fig. 9.8 and Fig. 9.9, respectively.

The operating state monitoring module recognizes the current operating state based on the flow rate and ion concentrations of the feeding solution, ORPs of each reactor, etc. The parameter monitoring and alarm module monitors the key process variables. When a key process variable exceeds its reasonable limitation or a device/equipment fault is detected or predicted, an alarm will be generated to alert the operators.

By using the fault and alarm function of ICS, the operators can respond before the fault can further generate large fluctuations of the purification process. As the technical indices/manipulated variables are optimized according to the technical features and operating conditions of the purification process, the application of the ICS provides more reasonable operations which improves the stability and efficiency of the process.

Chapter 10

Conclusions and future research

Contents

10.1 Summary

In this book, the authors present modeling, optimization, and control strategies for the zinc hydrometallurgical purification process. The overall research work follows the classical design procedure for the operational optimization of a nonferrous metallurgical process, which mainly includes the following steps.

Step 1. Process analysis: Obtain a deep and comprehensive understanding of the production technology and reaction kinetics. Determine the reaction type, and build the kinetic model of the process. Summarize the characteristics of the production process.

Step 2. Control problem analysis: Determine the control objectives, control inputs, and controlled variables. Confirm the physical limitations of control inputs and controlled variables. Analyze the difficulties in achieving the control objectives.

Step 3. Control problem formulation: Design indicators representing the technical and economical performance based on the characteristics of the production process. Then use the designed indicators to abstract and formulate the control problem.

Step 4. Control framework design: Design an optimization and control framework according to the characteristics of the production process and the control problem formulation. Obtain the solution for each module of the framework, e.g., optimization strategy, KPI controller, process monitoring, etc.

Step 5. Control system development: Develop a reliable yet human-friendly control system which realizes the functions of the control framework.

Step 6. Implementation, evaluation, and maintenance: Install the control system. Evaluate the control performance and make appropriate corrections to iteratively improve the control system.

https://doi.org/10.1016/B978-0-12-819592-5.00022-3

From the perspective of academic research, the key steps in the above procedure are Steps 1–4, which vary from process to process. As for the zinc hydrometallurgical purification process introduced in this book, the main results corresponding to Steps 1–4 include:

(1) The competitive-consecutive reaction system: As introduced in Chapter 3, multiple parallel chemical reactions occur in the copper removal process. These chemical reactions have competition or promotion relationships. Therefore, a modeling approach based on a competitive-consecutive reaction system was proposed for the copper removal process, which can also be applied in other processes with multiple parallel "competitive-consecutive" chemical reactions.

(2) The "ORP–reaction rate" relationship: Multiple parallel electrode reactions occur in the cobalt removal process. These reactions have competitive-consecutive relations, but share a common ORP. As these reactions involve transfer of electrons between the anode and the cathode, ORP determines the electron transfer rate (reaction rate) by affecting the potential difference. The discovery of this relationship facilitates the control framework design.

(3) Gradient optimization of cascaded reactors: In process industry, many unit processes are composed of cascaded reactors. The optimization problem of this type of process is essentially the cooperative optimization problem of each reactor. The gradient optimization approach introduced in this book transforms the cooperative optimization problem of the cascaded reactors to optimizing the variation gradient of KPIs along the reactors.

(4) The controllable domain-based fuzzy control approach: The attainable control performance varies with working conditions. In this book, the concept of controllable domain is introduced, which is then classified according to their attainable control performance. Fuzzy control rules are extracted for different controllable domains. By classification of the controllable domain, one can select appropriate control rules for the current working conditions rather than low-efficiency or high-cost control rules. This approach can be applied to other processes with varying working conditions.

The above results are obtained by an in-depth kinetic research, a long-period field study, and the application of advanced process control approaches. These results provide solutions for the modeling, optimization, and control problem of the solution purification process and support the operational optimization of the solution purification process.

10.2 Future research directions

Since the last decade, driven by the rapid development of ICT technologies, the mode of industrial automation is trending towards smart and optimal manufacturing [1][2][3]. The main idea behind this trend is essentially to globally optimize the allocation of production resources and achieve high-efficiency production by integration of ICT technologies and control theories on various levels. These new technologies can be roughly classified into two categories, including infrastructure-oriented and analysis-oriented technologies. The infrastructure-oriented technologies include cloud computing, edge computing, 5G communication, Industrial Internet of Things, etc. These approaches are aimed to gather, organize, and present the plant-wide data or the business data of the whole supply chain. The analysis-oriented technologies include big data analysis, transfer learning, reinforcement learning, deep learning, etc. These approaches are aimed to take optimization and control actions based on useful information/knowledge extracted from the collected data and the discovered reaction kinetics, including:

(i) global optimization of the horizonal value chain,
(ii) cooperative optimization and intelligent control of the vertical production process.

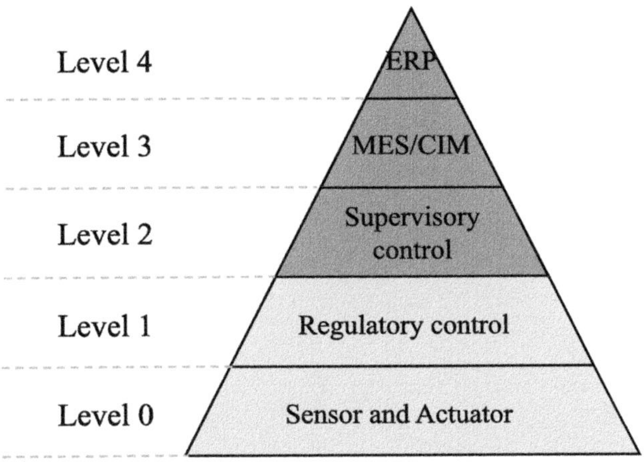

FIGURE 10.1 PERA pyramid structure of an industrial automation system.

Driven by this trend, the traditional pyramid structure of automation systems of process industries defined in the Purdue Enterprise Reference Architecture (PERA) (Fig. 10.1) is becoming more flat (Fig. 10.2). The intelligence originally only located at the supervisory level is also moving to the floor level. As a result, each single reactor, which is the unit material processing unit, is becoming more smart and able to make control decisions autonomously. The originally blocked lateral information flow between reactor/unit processes is unblocked to enable cooperation. Therefore, a single reactor can be viewed as an industrial agent or a cyber-physical component of a plant-wide cyber-physical system (Figs. 10.2 and 10.3). The above changes introduce new research topics. On the one hand, "plant-wide level cooperation" and "reactor-level autonomous control" are two new research problems of process industries. On the other hand, there are some existing common modeling/optimization/control problems of process industries

the change of working conditions or set-point. In order to achieve this aim, the autonomous controller should be able to:

(i) Know how the process behaves: Have a deep and comprehensive understanding of the dynamics of the reactions conducted in a reactor. The behavior of a process has many facets. Describing the process behavior relies not only on an in-depth study of the reaction kinetics but also on the extraction of reaction laws from production data using machine learning methods. According to the authors' previous study, both first-principle modeling and data-driven input/output modeling are informed by probability theory and statistics, e.g., the conservative laws and reaction kinetics used to construct the first-principle model are discovered from experimental data. The combination of these two methods could increase the utilization rate of data, information, and knowledge, thus leading to the discovery of more facets of the process dynamics. In addition, new facts about the process behavior can be discovered by inferencing from existing knowledge about the process dynamics. Therefore, integrating the reaction kinetic model, machine learning algorithms, and casual inference methods in an intelligent perception framework is a future research direction. Moreover, integration of reaction kinetics and production data requires a descriptive system which can also support the digitalization and visualization of the process. The comprehensive state space descriptive system proposed by the authors is a potential solution to address this problem.

(ii) Know how the working conditions are changing: For each reactor, its control task is to adjust the manipulated variables to force the technical indices to follow their set-points. However, under different working conditions, the behaviors of the reactor are different. In addition, the control tasks are also different. For example, for a cobalt removal reactor, if the inlet copper ion concentration is increased, the zinc powder dosage should be correspondingly increased to account for the working condition change. The increment of zinc powder dosage is determined by various influencing factors, e.g., inlet flow rate, flow rate of spent acid and arsenic trioxide, etc. The working conditions are determined by both inlet conditions and reaction conditions. In practical industrial processes, changes of working conditions are mainly caused by changes of inlet conditions, which include basic measurements like flow rate and discrete-delayed measurements like concentration. If the reaction conditions and the manipulated variables are not changed appropriately, then the working conditions may deteriorate. In order to obtain their real-time value, the discrete-delayed measurements need to be estimated. Therefore, the estimation of inlet conditions is an important step in knowing how the working conditions are changing. Theoretically, the estimation of inlet conditions is achieved by designing a state observer which reconstructs the inlet conditions from process measurements [4]. With the rapid development of deep learning algorithms, the main concern in designing an observer for inlet conditions is not the approximation ability of the algorithm but the design of features systematically describing the inlet conditions and the structural design of an estimation framework which has a high utilization rate of measurement/information/knowledge and low risk of large estimation error. The estimation of inlet conditions is an "old" topic, however, deserving further study using "new" approaches.

(iii) Know how to adjust the manipulated variables: A reactor has various working conditions which can be classified roughly as normal working conditions and faults [5][6]. Under different working conditions, different control strategies (or control rules) should be applied. In order to make appropriate selection under different working conditions, these different control strategies should be incorporated in a framework. A fault detection and diagnostics (FDD) unit is placed over the control layer to monitor the working conditions online. On the one hand, when a fault occurs, the FDD unit outputs the type of fault. Then, the self-recovery unit adjusts the configuration of the control framework according to the effectiveness of each control loop/path, the design objective function for self-recovery, and derive reasonable control actions. On the other hand, if the process is under normal working conditions, then optimal control should be derived to obtain the optimal control performance. Faults and normal working conditions are only rough classifications. The normal working conditions can be further divided into various working conditions under which the model parameters take different values. Moreover, under some working conditions, the model parameters are hard to obtain due to reasons like limited data samples for identification or the working condition is a new one. As a result, the model-based control approaches cannot be directly applied under these situations. Therefore, model-based and model-free control approaches should be integrated to cover all the normal working conditions. The design of an appropriate framework which incorporates both model-based and model-free control approaches and enables the interaction between the two types of control approaches is required. Please note that rule-based control approaches can also be applied in this scenario. Then the intelligent extraction, selection, and evolution of rules under different working conditions become the research topic. In addition, machine learning approaches can improve the efficiency of optimization and control approaches by learning from production data [7][8]. Therefore, the integration of advanced control, FDD, self-recovery, and machine learning approaches can improve the production efficiency and economic performance by providing more accurate control actions.

10.2.2 Plant-wide intelligent cooperation

An entire plant is composed of associated unit processes via energy/mass transfer. Plant-wide optimization is an important feature of smart and optimal manufacturing. It considers the optimization problem on the scale of the whole industrial production chain, which can achieve global optimization of associated unit processes rather than local optimization of a unit process. As the dynamics of a plant are nonlinear, the technical indices of each unit process exhibit complex relationships. The plant-wide optimization problem is difficult to formulate and solve explicitly. As an alternative solution, the plant-wide optimization can be approximated by finding the optimal cooperation pattern of associated unit processes, or in other words, finding the optimal operation mode of each unit process.

The entire plant can be viewed as a network of unit processes composed of cascaded reactors. Each reactor capable of autonomous control can be considered as an industrial agent. The entire plant, unit process, and reactor form a three-level system. In order to establish a plant-wide intelligent cooperation framework, three problems need to be studied:

(i) Know how to describe the interactions between unit processes/reactors: There exist interactions between unit processes and reactors due to energy/mass transfer. Due to these interactions, the unit processes or reactors are not isolated. Reasonable cooperations between unit processes/reactors can damp the fluctuations of feeding conditions, guarantee the stability of product quality, and optimize the overall production cost. As a foundation, the plant-wide intelligent cooperation framework should be capable of describing the interactions between unit processes/reactors. As discussed above, an entire plant could be viewed as a material flow and processing network in which a unit process or a reactor acts as a node. The graph network, which is composed of nodes and edges, is a natural approach to describe the relationship between unit processes or reactors. These relationships could be extracted by applying machine learning approaches and kinetic models. Therefore, combining reaction kinetics and machine learning approaches together within a graph network is a potential research topic.

(ii) Know how to find the optimal plant-wide cooperation pattern: Due to the internal interactions between unit processes/reactors, if a unit process is in a different operating mode, then it influences its associated unit processes differently. So there exists an optimal operating mode for each unit process which forms an optimal plant-wide cooperation pattern. Moreover, when the feeding conditions and working conditions change, the operating mode of each unit process should be adapted to form a new optimal plant-wide cooperation pattern. The determination of the optimal plant-wide cooperation pattern belongs to the realm of decision making which can be achieved by knowledge-based inferencing on a graph network. Knowledge automation, which involves knowledge representation, acquisition, association, processing, and application, is a potential solution for this problem [9].

(iii) Know how to transfer between cooperation patterns: After the optimal plant-wide cooperation pattern for the current feeding conditions and working conditions are determined, the set-point of technical indices of each unit process and reactor have to be obtained to enable the transfer of cooperation patterns. The cooperation pattern transfer problem belongs to the realm of operational optimization or optimal control if the technical indices are taken as control inputs. The optimization objectives include the stability, cost, and final offset of the transfer process. The optimization problem involves deriving the optimal trajectories of various variables. Deep reinforcement learning, which combines the precise perception ability of deep learning and the optimal control ability of reinforcement learning and can learn the optimal control sequence from previous operation experiences, could be an effective approach [10].

10.2.3 Epilogue

The process industry is experiencing a transformation from the era of advanced process control to the era of smart and optimal manufacturing. Besides the improvement of infrastructure which facilitates smart and optimal manufacturing, the rapid development of machine learning technologies as well as learning-based control approaches and their applications in process industries enable the increase of intelligence of the entire plant. This does not mean that the study of reaction kinetics is no longer important. Instead, only by a deep integration of reaction kinetics, control theory and machine learning theory can pave the way from data, information, and knowledge to intelligence. As the fundamental theories and key technologies for smart and optimal manufacturing cover the horizonal supply chain and the vertical production cycle, the future research directions listed in this section are far from complete. Various relevant scientific problems need to be studied before the fundamental theories and key technologies of smart and optimal manufacturing can be systematically established, which is the key to achieve a higher level of green and efficient production.

References

[1] D. Zuehlke, SmartFactory towards a factory-of-things, Annual Reviews in Control 34 (1) (2010) 129–138.

[2] T. Chai, Industrial process control systems: research status and development direction, SCIENTIA SINICA Informationis 46 (8) (2016) 1003–1015.

[3] F. Qian, W. Zhong, W. Du, Fundamental theories and key technologies for smart and optimal manufacturing in the process industry, Engineering 3 (2017) 154–160.

[4] J. Mohd Ali, N. Ha Hoang, M.A. Hussain, D. Dochain, Review and classification of recent observers applied in chemical process systems, Computers & Chemical Engineering 76 (2015) 27–41.

[5] Y. Du, H. Budman, T.A. Duever, Integration of fault diagnosis and control based on a trade-off between fault detectability and closed loop performance, Journal of Process Control 38 (2016) 42–53.

[6] L.E. Olivier, I.K. Craig, Should I shut down my processing plant? - An analysis in the presence of faults, Journal of Process Control 56 (2017) 35–47.

[7] B.R. Haverkort, A. Zimmermann, Smart industry: how ICT will change the game!, IEEE Internet Computing 21 (1) (2017) 8–10.

[8] C. Shang, F. You, Data analytics and machine learning for smart process manufacturing: recent advances and perspectives in the big data era, Engineering 5 (2019) 1010–1016.
[9] W. Gui, X. Chen, C. Yang, Y. Xie, Knowledge automation and its industrial application, SCIENTIA SINICA Informationis 46 (8) (2016) 1016–1034.
[10] V. Mnih, K. Kavukcuoglu, D. Silver, A.A. Rusu, J. Veness, M.G. Bellemare, A. Graves, M. Riedmiller, A.K. Fidjeland, G. Ostrovski, et al., Human-level control through deep reinforcement learning, Nature 518 (7540) (2015) 529–533.

Index

D

E

F

G

H

I

O

P

Z

CPI Antony Rowe
Eastbourne, UK
June 11, 2021